CONSTRUCTION OF CONCRETE BREAKWATER AT LORAIN, OHIO.

A coal-loading and ore-unloading port on Lake Erie.

Cement Construction

Radford Architectural Company

Fredonia Books
Amsterdam, The Netherlands

Cement Construction

by
Radford Architectural Company

ISBN: 1-4101-0602-0

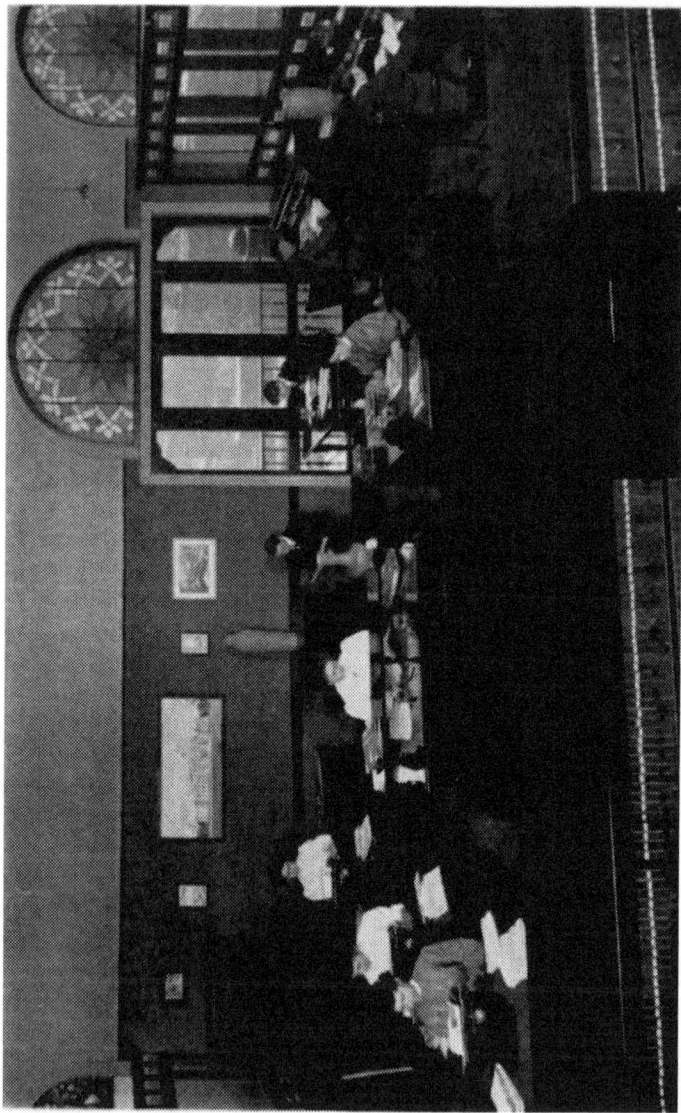

CYCLOPEDIA DEPARTMENT, THE RADFORD ARCHITECTURAL COMPANY, CHICAGO, ILL.

Preface

THE Building Industry, in its various branches, is more closely identified than any other with the marvelous engineering progress of the present day and the untold possibilities of the future. To put all classes of workers in familiar touch with modern methods of construction and the latest advances in this great field, and to bring to them in a form easily available for practical use the best fruits of the highest technical training and achievement, is the service which the CYCLOPEDIA OF CONSTRUCTION aims to render.

The work is pre-eminently a product of practical experience, designed for practical workers. It is based on the idea that even in the larger problems of engineering construction, it is not now necessary for the ordinary worker in concrete, or steel, or any other form of material, to attempt the impracticable task of exploring all the highways and byways where the trained engineer or technical expert finds himself at home. The theories have been worked out; the tests and calculations have been made; observations have been recorded in thousands of instances of actual construction; and the results thus accumulated form a vast treasure of labor-saving information which is now available in the shape of practical working rules, tables, instructions, etc., covering every phase of every construction problem likely to be met with in ordinary experience. This is perhaps most apparent in the sections on *Cement and Concrete Construction, Plain and Reinforced*. To this subject, on account of its supreme importance as a structural factor of the present day, three entire volumes are devoted, embodying the cream of all the valuable information which engineers have gathered up to date. Much of this practical information now presented in this Cyclopedia, has never before been published in any form. By its use,

PREFACE

anyone is enabled to take advantage of the vast labors of others, and to bring to bear on any problem confronting him the results of the widest experience and the highest skill.

The keynote of the Cyclopedia is found in the emphasis constantly laid on the *practical* as distinguished from the *theoretical* form of treatment, in its total avoidance of the complicated formulas of higher mathematics, and in its reduction of all technical subjects to terms of the simplest and clearest English. Throughout the pages devoted to Steel Construction, for example, the mathematics of the subject have been eliminated to such an extent that *the reader will not find a single instance where even a square root sign has been used.*

In addition to the larger problems of engineering and building construction, one entire volume, as well as many chapters scattered through the work, is devoted to those smaller constructions that are of special interest to the teacher or student of manual training or the home shop worker of a mechanical turn of mind.

Inasmuch as a wider knowledge and a more intelligent grasp of the fundamental principles of construction and design will tend to greater efficiency on the part of working-men, and to greater economy in production, the purpose of the CYCLOPEDIA OF CONSTRUCTION is one which will appeal strongly, not only to the men themselves, but also to the architectural and engineering fraternity as a whole.

The authors of the various sections are all men of wide experience whose recognized standing is a guarantee of reliability and practical thoroughness.

Table of Contents

PLAIN CONCRETE CONSTRUCTION . . . PAGE 1
Why Build with Cement?—Cost and Design of Forms—Time
to Remove Forms—Causes of Failure—Color Variations—
Proper Finish—Collapsible Metal Forms (Miracle, Overturf,
Blaw)—Metal Forms for Building Construction (Monolith,
Jackson) — Wall Construction — House Foundations — Fin-
ishes for Concrete Surfaces (Bush-Hammering, Tooling,
Acid Treatment, etc.)—Stucco—Colored Aggregates—Metal
Lath—Asbestos Lumber—Concrete Roofs—Cement Shingles
and Roofing Tile—Cement Brick for Chimneys—Asbestos
Shingles—Concrete Floors and Steps—Sarco Mastic—Veranda
and Cellar Floors—Jointless Floors—Flying Stairs—Concrete
Piazza—Chimney Caps—Fireplace—English Systems of Floor
Construction—Fireproofing—Systems of House Construction
(Edison Poured House, Graham System, Hollow-Wall Method,
Aiken Method)—Blocks without Facing—Color of Blocks—
Coloring of Mortar—Block Industry—Manufacture of Blocks

CONCRETE ON THE FARM . . . . PAGE 108
Windmill Foundation—Sinks—Water-Tanks—Cisterns—Well-
Curbs—Silos (Hollow-Wall, Monolithic Solid-Wall, Block)—
Sizes and Feeding Capacities of Silos—Barn and Stable
Floors and Foundations—Feeding Floors—Box Stalls—Piers
and Posts—Rain Barrel—Horse-Block—Hotbed Frames—
Greenhouses — Fountain — Road Culverts — Sewer Pipe—
Chicken House—Hen's Nests—Hog Pens—Ice-House—Stor-
age Buildings—Root Cellar—Cyclone Cellar—Mushroom Cel-
lar—Fence and Gate Posts—Ornamental Shapes

SIDEWALK CONSTRUCTION . . . . PAGE 160
Materials Used—Proportions and Mixing—Sub-Base—Drain-
age—Tools—Forms—Expansion Joints—Surface Treatment—
Data for Estimating—Causes of Defects—Curbs and Gutters
—Protection of Corners—Drain Tile

REINFORCED CONCRETE CONSTRUCTION . . PAGE 183
Fundamental Principles and Methods—Tensile, Compressive
and Shearing Forces—Historical Sketch—Advantages of Re-
inforced Concrete—Cost—Wood and Concrete Piles—Fire Risk
and Insurance—General Principles of Design—Neutral Axis
—Reinforcing Steel (Bars, Unit-Frames, Structural Shapes,
Sheet Fabrics)—Simple Practical Rules for Design (Wason's
and Ransome's Formulas)—Design of Columns—Materials
Used for Floors and Roofs—Allowable Stresses—Building
Laws of Various Cities

REINFORCING MATERIALS AND SYSTEMS . PAGE 238
Steel Bars (Plain, Deformed, Trussed)—Expanded and Rib
Metal—Sheet Metal—Steel Lath—Wire Fabric—Steel Shapes
—Bending and Twisting—Life of Steel in Concrete—Systems
of Reinforcement (Ransome, Kahn, Gabriel, Columbian, Cummings, Mushroom, Spider-Web, Cowles Umbrella, Unit-Beam,
Unit-Girder Frame, System M, De Vallière, Hennebique, Merrick, Roebling, "Standard," Siegwart, "American" High-Carbon, Vaughan, National, etc.)—Column Footings—Concrete
Piles (Simplex, Raymond, Gow, Gilbreth, Corrugated, Chenoweth, Hennebique, etc.)

GENERAL BUILDING CONSTRUCTION . . PAGE 296
Frame and Steel Construction—Mill or Slow-Burning Construction—Erection—Beam and Column Forms—Mixtures
Adaptable to Various Classes of Work—Floors, Slabs, and
Roofs—Flat and Arched Slabs—Commercial Systems—Girderless Floor—Columns (Kahn, Smith Spiral, Lally, Hennebique,
Mushroom, etc.)—Terra-Cotta Tile—Reinforced Walls—Ceilings and Partitions—Finishes for Floors and Walls—Sizes
and Gauges of Expanded Metal

INDEX . . . . PAGE 331

CONCRETE CULVERT IN PROCESS OF CONSTRUCTION.

One abutment finished. View showing men mixing concrete and filling form for other abutment.

Cement Construction

WHY YOU SHOULD BUILD WITH CEMENT

The imperishability and the artistic qualities of a well-built cement structure are unquestioned. As to evidence of the durability of this wonderful material, we have only to visit some of the ancient buildings of the Eternal City to see a verification of its lasting quality. Here are the principal reasons why you should build with cement:

A cement building will last practically forever.
It can be made fireproof.
It is warmer in winter and cooler in summer.
It requires no paint and no repairs.
It is earthquake-proof and defies the elements.
It is cheaper than any other form of construction.

In constructive work the primal factor is durability outside of all considerations of beauty and form. Buildings or walls made of cement are independent of the painter. No patching has to be done. All is imperishable. The home builder is awakening to the truth of these facts and the architect is adapting himself to the new conditions. Mr. Moyer has said:

"If you employ concrete, let it look like concrete, design for concrete, eliminate all thought of stone, brick,

1

wood, or plaster. Let the house stand up and be able to say to the casual observer, 'I am solid, strong, substantial, durable, beautiful, and constructed of concrete.' That which looks right to the practiced and trained eye is right. For country residences, particularly where there are winding roads, trees, a hillside, and possibly rocks, concrete treated as concrete looks right."

FORM CONSTRUCTION

A Field for Invention. Concrete, being a plastic material and requiring time to set, must needs be confined, during the chemical process of setting, within the bounds and in the shape and form the completed structure or member is to possess. The **form** or temporary support that so holds the material during this setting process, must be rigid. The materials commonly used are wood and iron. In building construction, the use of wooden forms may be said to have been hitherto the almost universal practice, though several types of metal forms have been invented and are now on the market.

There is a great field for research and inventive genius in the production of cheap, durable **forms** sufficiently flexible for many different uses, as the cost of this part of concrete work is at present one of the greatest handicaps in the way of a reduction of cost and a more general use of concrete construction.

The writer has used metal forms to some extent, and has observed their use by contractors. If sheet metal is placed on a wooden back or on a metal stiffening frame, there is danger of its becoming dented, bent or otherwise defaced so as to give an imperfect surface to the concrete; and if the metal covering is sufficiently thick and strong to resist damage, it is likely to be too heavy and expensive for general use. Moreover, such forms are not flexible so that they can be used for various purposes. Wood has thus far, therefore, proved to be the most economical and flexible in the way of being changed from one use to another of anything which the writer has seen, and his study has been concentrated on the most effective and economical use of lumber. When a sheet metal form becomes dented, it is usually cheaper to throw the covering away and start new than to straighten and re-use that which is bent.

The cost of construction as a whole is as likely to be governed by the cost of the forms as the cost of the concrete. For illustration, in the cost of columns, although they are made so as to be reduced in size as easily as possible, this is somewhat expensive in labor; and after they are reduced, the girders and beams which meet at a column are too short and have to be spliced out, which adds to the cost, and these costs are likely to exceed that which can be saved in concrete. It is therefore more economical to run columns one size through the full height of a low building, or to reduce the size only twice, or at most three times, in the height of a high building. For instance, the actual cost of labor (without regard to wastage of lumber) in reducing columns from 16 inches to 12 inches on a certain job amounted to $5.70, whereas the cost of the concrete saved by the reduction was only $2.30. In the writer's office a set of plans in pamphlet form has been compiled for standard forms for all kinds of work; and where special framing is required, plans are made for these special cases. But in spite of this care and study, the above figures represent actual experience. Therefore, in designing and handling form work, the **cost of labor** is the principal item to be considered.

The opinion was expressed by Mr. Larned that the forms could be cheapened by the use of common and rough lumber instead of a fairly good grade of dressed lumber. The fact that it is almost universal to use a good grade of dressed stock would seem to disprove the above statement. The cost of dressing varies, according to the mill, from $1.50 to $4.00 per thousand feet for planing four sides. The price of second-grade hemlock and spruce differs but little from that which has small, sound knots and is free from wind-shakes or large season cracks; and such lumber works so much easier as to cut down the labor cost more than the equivalent of the difference in cost of material.

Some builders use $\frac{7}{8}$-inch stock; others, $1\frac{1}{2}$-inch; and some, 2 inches thick. The thicker lumber will stand the wear and tear longer than the light, and can therefore be used so many more times than the thin that it is more economical in the long run if the work in hand is large enough so that the forms can be used several times without delaying the rapid progress of the work. With planed stock, tighter joints can be obtained between boards, which prevents leakage of the fine materials and weakening and roughening of the surface; and the boards are of even thickness, so that a fairly good surface is obtained which needs little treatment after the forms are removed, except where an ornamental appearance is especially desired. If rough lumber is used, in order to get even a passable surface finish, considerable labor must be spent upon dressing the concrete after the forms are removed, and this must be done by mechanics.

Design of Forms. "Rule-of-thumb" layout of forms in the field is being superseded by design in the drafting-room. In building construction where the forms form a large percentage of the cost of the building, and where a failure in the forms may cause loss of life, it is especially necessary to treat this question from an engineering standpoint, and many of the best concrete contractors now design their forms as carefully as the dimensions of the concrete members.

If a minimum quantity of lumber is to be used consistent with the deformation allowed, it follows that the dimensions and spacing of the supporting lumber must be actually computed from the weight or pressure against the sheeting. For columns and for walls where a considerable height of wet concrete is to be placed at once, the pressure may be calculated as a liquid. Mr. W. J. Douglas assumes that the concrete is a liquid of half its own weight, or 75 pounds per cubic foot.

In ordinary walls, where the concrete is placed in layers, computation is not usually necessary, since general experience has shown that maximum spacing for 1-inch boards is 2 feet, for 1½-inch plank is 4 feet, and for 2-inch plank is 5 feet. Studding generally varies from 3 by 4 inch to 4 by 6 inch, according to the character of the work and the distance between the horizontal braces or waling, 4 by 4 inch being the most useful size.

Floor forms are better based upon an allowable deflection than upon strength, in order to give sufficient stiffness to prevent partial rupture of the concrete or sagging beams.

In calculating we must add to the weight of the concrete itself—that is, to the dead load—a construction live load which may be assumed as liable to come upon the concrete while setting. Definite units of stress must also be assumed in the lumber.

We would suggest the following basis for computation, these being values which have been adopted for use:

(1) Weight of concrete, including reinforcement, 154 lbs. per cu. ft.

(2) Live load, 75 lbs. per sq. ft. upon slab, or 50 lbs. per sq. ft. in figuring beam and girder forms.

(3) For allowable compression in struts, use 600 to 1,200 lbs. per sq. in., varying with the ratio of the size of the strut to its length. (See table below.) If timber beams are calculated for strength, use 750 lbs. per sq. in. extreme transverse fiber stress.

(4) Compute plank joists and timber beams by the following formula, allowing a maximum deflection of ⅛ inch:

$$d = \frac{3Wl}{384EI} \dots \dots (1)$$

and,

$$I = \frac{bh^3}{12} \dots \dots (2)$$

in which,

    d=Greatest deflection in inches;
    W=Total load on plank or timber;
    l=Distance between supports in inches;
    E=Modulus of elasticity of lumber used;
    I=Moment of inertia of cross-section of plank or joist;
    b=Breadth of lumber;
    h=Depth of lumber.

The formula is the ordinary formula for calculating deflection except that the coefficient is taken as an approximate mean between $^1/_{384}$ for a beam with fixed ends, and $^5/_{384}$ for a beam with ends simply supported.

For spruce lumber and other woods commonly used in form construction, **E** may be assumed as 1,300,000 lbs. per sq. in.

Formula (1) may be solved for **I**, from which the size of joist required may be readily estimated.

The **weight of concrete** per cubic foot is somewhat higher than is frequently used, but is none too much where a dense mixture and an ordinary percentage of steel is used. For very rough calculation, however, it is frequently convenient to remember that 144 lbs. per cubic foot is equivalent to the product of the dimensions of the beam in inches times a length of one foot.

The suggested live load is assumed to include the weight of men and barrows filled with concrete, and of structural material which may be piled upon the floor, not including, however, the weight of piles of cement or sand or stone, which should never be allowed upon a floor unless it is supported by concrete sufficiently strong to bear the weight, or by struts under all the floors below.

The units for stress in struts are somewhat higher than in timber construction, because the load is a temporary one. The extreme variation given is due to the fact that when a column or strut is longer than about sixteen times its smallest width there is a tendency to bend, which must be prevented either by bracing it both ways or by allowing a smaller load per square inch. For struts ordinarily used, the stresses given in Table I may be assumed for different heights.

Bracing both ways will, of course, reduce the length of a long strut.

If the concrete floor is comparatively green, the load must be distributed by blocking, preferably of hardwood. At the top of the strut, pro-

TABLE I

Safe Strength of Wood Struts in Forms for Floor Construction.

POUNDS PER SQ. IN. OF CROSS-SECTION.

LENGTH OF STRUT.	3" x 4"	4" x 4"	6" x 6"	8" x 8"
14 ft.		700	900	1,100
12 ft.	600	800	1,000	1,200
10 ft.	700	900	1,100	1,200
8 ft.	850	1,050	1,200	1,200
6 ft.	1,000	1,200	1,200	1,200

vision must be made against crushing of the wood of the plank or cross-piece. Ordinary soft wood will stand without crushing only about 700 lbs. per sq. in. across the grain; so, if the compression approaches this figure, brackets must be inserted or hardwood cleats used.

Kinds of Lumber to Use. The selection of the lumber must be governed by the character of the work and the local market. White pine is best for fine facework, and quite essential for ornamental construction cast in wooden forms. For ordinary work, however, even for the panels, white pine is apt to be too expensive; and spruce, fir, Norway pine, or the softer qualities of Southern pine must be substituted for it. Some of these woods are more liable to warp than white pine, but they are generally stiffer, and thus better adapted for struts and braces.

Kiln-dried lumber is not suitable for form construction, because of its tendency to swell when the wet concrete touches it. Very green lumber, on the other hand—especially Southern pine, which does not close up quickly when wet —may give trouble by joints opening. Therefore the middle ground—or, in other words, partially dry stuff—is usually best.

Finish and Thickness of Lumber. Either tongued-and-grooved or bevel-edged stuff will give good results for floor and wall panel forms, and is preferable to square-edged stuff. A smoother surface may be attained at first with the tongued-and-grooved stock than with square- or bevel-edge, and there is less trouble with opening joints, but it is more expensive because of the waste in dressing; and if the forms are used many times, there is greater tendency to wear at the joints. Even for rough forms, plank planed one side may be economical to cheapen the cost of cleaning. Studs should always be planed one side to bring to size.

The thickness of lumber varies with different contractors, some using 1-in., others 1½-in., while a few employ 2-in. stuff even for panels, these being commercial thicknesses measured before planing. For ordinary walls, 1½-in. stuff is good; although, for heavy construction where derricks are used, 2-in. is preferable; while for small panels, 1-in. boards are lighter and easier to handle. For floor panels, 1-in. boards are most common; although, if the building is eight stories high or over, 1-in. stuff of soft wood is likely to be pretty well worn out before the top of the building is reached, and the under surface of the concrete will show the wear badly. For sides of girders, either 1-in. or 1½-in. is sufficient, while 2-in. is preferable for the bottoms of girders. Column forms are generally made of 2-in. plank.

In building forms and centering, remember

that the cost of timber is a large item of the total cost. With a little ingenuity and forethought, this work can be most always arranged so that repeated use is made of each piece of timber.

Alignment of Forms. Alignment is another item of importance, since it is here that a great deal of time may be wasted by inexperienced or incompetent carpenters. Such workmen may err either on the side of poor alignment or more careful alignment than the structure requires. Mr. W. J. Douglas suggests as a general rule the allowance of ⅜-inch departure from established lines on finished work, and 2 inches on unfinished work.

In removing forms, the green concrete must not be disturbed by prying against it. This seems so obvious as to need no emphasis; but I have known first-class carpenters to attempt to straighten a wall which was an inch out of line, the day after the concrete was laid, by prying the forms over. The wall was straightened, but by a different process from that proposed by the carpenter—the concrete was relaid.

Forms for facework should be tightly put together, it being advisable in some cases to close the joints and holes by mortar, putty, plaster of Paris, sheathing paper, or thin metal. This is not, as is commonly supposed, to prevent loss of strength by the cement which flows out with the water, but rather to prevent the formation of voids or stone pockets in the finished surface.

Use of Oil. Crude oil is one of the best materials to prevent adhesion of the concrete to forms, though linseed oil, soft soap, and various other greasy substances are also employed for this purpose. The oil or grease should be thin enough to flow and fill the grain of the wood.

If the forms are to be left on until the concrete is hard, there is little danger of the concrete sticking to them if they are wet thor-

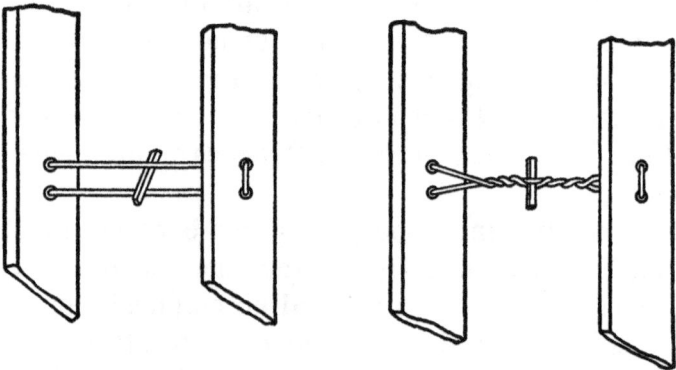

Fig. 1.   Tightening and Steadying Forms by Twisted Wire Loops.

oughly with water before the concrete is laid instead of being greased. In any case, if concrete adheres to the forms it should be thoroughly cleaned off before resetting; even then it is apt to stick again in the same place.

Protecting the Forms. After the falsework is once placed, it should be protected against severe temperature changes, and against rain or snow. If it is necessary to leave the form work empty for any length of time, then, before pouring the concrete, it should be gone over again carefully, the shrinkage joints being repaired and the bracing and supports tightened up. In warm weather, wet the forms before placing concrete; otherwise the lumber will absorb the water in the concrete, and cause honeycombing. For form boxes of columns or beams, 2-inch timber is recommended; for floor slabs, 1-inch boards on 2 by 8-inch joists, spaced 16 to 24 inches apart, according to thickness of slab; for posts, 4 by 4 inches spaced from 3 to 8 feet apart, according to the height and the dimensions of beams and slabs; for bracing and cross-bracing, 2 by 4-inch.

The skilful superintendent or foreman may be recognized by the scarcity of nails he uses. Column boxes should be securely clamped, and beam boxes so framed that the floor joists practically hold the sides without nailing. If the removal of centering necessitates the use of the long prying crowbar with three laborers at its end, the foreman should be fired.

There should be one foreman, at least, for every twenty carpenters, and a head carpenter foreman in charge of all falsework.

Time to Move After Placing. The best contractors have definite rules for the minimum time the forms must be left in ordinary weather; and then these times are lengthened for changes in conditions according to the judgment of the foreman.

Conference with a number of prominent contractors in various parts of the country indicates substantial agreement in the minimum time to leave forms. As a guide to pratice, the following rules are suggested, these following in the main the requirements of one large construction company:

Walls in mass work—one to three days, or until the concrete will bear pressure of the thumb without indentation.

Thin walls—in summer, two days; in cold weather, five days.

Slabs up to 6 feet span—in summer, six days; in cold weather, two weeks.

Beams and girders and long-span slabs—in summer, ten days or two weeks; in cold weather, three weeks to one month. If shores are left without disturbing them, the time of removal of the sheeting in summer may be reduced to one week.

Column forms—in summer, two days; in cold weather, four days, provided girders are shored to prevent appreciable weight reaching columns.

Conduits—two or three days, provided there is not a heavy fill upon them.

Arches—of small size, one week; for large arches with heavy dead load, one month.

All of these times are, of course, simply approximate, the exact time varying with the temperature and moisture of the air and the character of the construction. Even in summer, during a damp, cloudy period, wall forms sometimes cannot be removed inside of five days, with other members in proportion. Occasionally, too, batches of concrete will set abnormally slowly, either because of slow-setting cement or because of impurities in the sand; and the foreman and inspector must watch very carefully to see that

CONCRETE HIGHWAY ARCH BRIDGE OVER ROCKY RIVER, CLEVELAND, OHIO.

Span of main arch, 280 ft., or over 47 ft. longer than that of the great Walnut Lane bridge at Philadelphia, Pa.; clear height above water, about 100 ft.; total length of bridge, 708 ft.; roadway, 40 ft. wide, with 8-ft. sidewalks at each side; cost, $20,000.

the forms are not removed too soon. Trial with a pick may assist in reaching a decision.

Beams and arches of long span must be supported for a longer time than short spans, because the dead load is proportionately large, and therefore the compression in the concrete is large even before the live load comes upon it.

The general uncertainty, and the personal element which enters into this item, emphasize the necessity for some more definite plan for insuring safety. The suggestion has been made that two or three times a day a sample of concrete be taken from the mixer and allowed to set on the ground under the same conditions as the construction until the date when the forms should be moved. These sample specimens may be then put in a testing machine to determine whether the actual strength of the concrete is sufficient to carry the dead and construction loads. Even this plan does not provide for the possibility of an occasional poor batch of concrete, so that watchfulness and good judgment must also be exercised.

Causes of Failure. Sanford E. Thompson, on **"Forms for Concrete Construction,"** in a National Cement Users' Association bulletin, says:

"Recent failures in reinforced concrete construction cannot be cast to one side and forgotten with the passing comment so frequently heard that the accident was due merely to poor construction or too early removal of forms. The reasons for every failure should be thoroughly investigated by experts to prevent recurrence of similar accidents.

"Forms, although frequently guilty, are by no means the only culprits. In fact they are frequently blamed when the designer is at fault. Just so long as men who know nothing of the first principles of mechanics are permitted to design concrete structures, and just so long as irresponsible contractors are engaged to erect them, the list of accidents will increase in startling numbers. In every case it is the men, not the inanimate lumber and materials, who are to blame. However, granting its danger in ignorant hands, reinforced concrete as a whole must not be condemned for failures due to improper conditions, any more than brick should be rejected as a building material for apartment houses, because of the collapse of several unfinished buildings in New York City a short time ago through disregard of frost action upon the mortar."

Failures in concrete buildings may be attributed to:

(1) Imperfect design, especially through neglect of essential details in locating the reinforcing metal, and through the adoption of too low a factor of safety.

(2) Poor materials, such as cement which does not properly set up, or sand which is too fine or which has an excess of clay, loam, or other impurities.

(3) Faulty construction, from improper proportioning, mixing, or placing, or too early removal of forms.

(4) Weak forms.

"A disregard of such important principles," says Mr. Thompson, "is frequently criminal negligence; and yet, in at least one case under my observation, an examination of the structure and the materials, after a collapse in which a number of lives were lost, showed the design, materials, and construction all faulty, so that it was impossible to decide positively which of the four causes named above was the primary reason for the failure.

"Certain general rules are applicable to all kinds of forms. **Strength, simplicity,** and **symmetry** are three fundamental principles of design. The necessity for strength is obvious, while economy in concrete construction consists in quickly erecting and moving the forms and in using them over and over again.

"The design of the concrete members should recognize the forms. A slight excess of concrete sometimes may be contributed to save carpenter work. Frequently beams may be designed of such widths as to use dimension widths of lumber without splitting.

"Columns may be of dimensions to avoid frequent re-making. Panel recesses in walls may be made the thickness of a board or plank. To permit ready cleaning of dirt and chips from the column forms before laying the concrete, at least one prominent contractor provides a door at the bottom of each of them.

"In building construction, the forms must be designed so that the column moulds and also the bottom of beam moulds are all independent of the slabs. The forms may thus be left a longer time upon members subjected to the greater stress.

"The sides of the beam moulds should be held tightly together by wedges or clamps, to prevent the pressure of the concrete springing them

away from the bottom boards. At top or bottom
of each strut, hardwood wedges are useful when
setting and removing it, and also permit testing,
to make sure that there is no deflection of the
beam or slab. For this purpose, some contractors loosen the wedges twenty-four hours in
advance of the struts. In general, it is preferable to use comparatively light joists, such as 2
by 8-inch or 2 by 10 inch, with frequent shores,
rather than to use lumber which is heavier to
handle.

"If forms are to be used but once or must be
taken apart when removed, it is sometimes practicable to use only a few partially driven nails so
that they can be withdrawn without injury to
the lumber. It is very difficult to convince house
carpenters that the pressure of the concrete will
hold temporary panel boards in place with
scarcely any nailing."

COLOR VARIATIONS

One of the difficulties encountered in plaster
and monolithic wall construction is a variation
of color between the various layers of concrete
that are deposited, and also the tendency in
monolithic walls of the aggregates and stones to
show on the surface, separating themselves from
the matrix, or cement. This variation of color
may be due to variation in the character of the
sand or to non-uniformity in proportions or mixing, or possibly to other causes. An artistic
method of overcoming these difficulties is to cut

the surfaces by grooves or channelings, giving the surface the effect of stone work. For this purpose the V-shaped form has been found to be best adapted. The V is placed in the form at intervals of eight or ten inches and when filled produces the familiar panel finish.

Proper Finish. Monolithic construction, to bring the cost of mill work for the forms within reasonable bounds, should be of simple design, involving no elaborate or intricate detail of mouldings or cornice effects. The use of the quarter-round is suggested, and all forms should be beveled so that they will pull away readily from the concrete without breaking off any edges. The proper finish for all exteriors of monolithic buildings is being given much attention by architects. It has been found that no matter how carefully the forms are made, when they are removed the wall has a cast appearance that is not desirable. A method frequently adopted to remove the pasty texture of the surface is to use a sand blast. This gives the surface a rough appearance, but in most cases the edges of the seams have to be dressed by hand.

Rubbing Down. Another finish has been obtained by oiling the forms and rubbing down the work after it is finished with carborundum blocks, and using cement paint, which is made with cement and water. The carborundum block cuts away the irregularities on the surface and the paint fills in the crevices, the two producing

smoothness. **Of course in the case of a very
large building this would be a laborious process,**
but for a residence it is not a difficult or expensive operation.

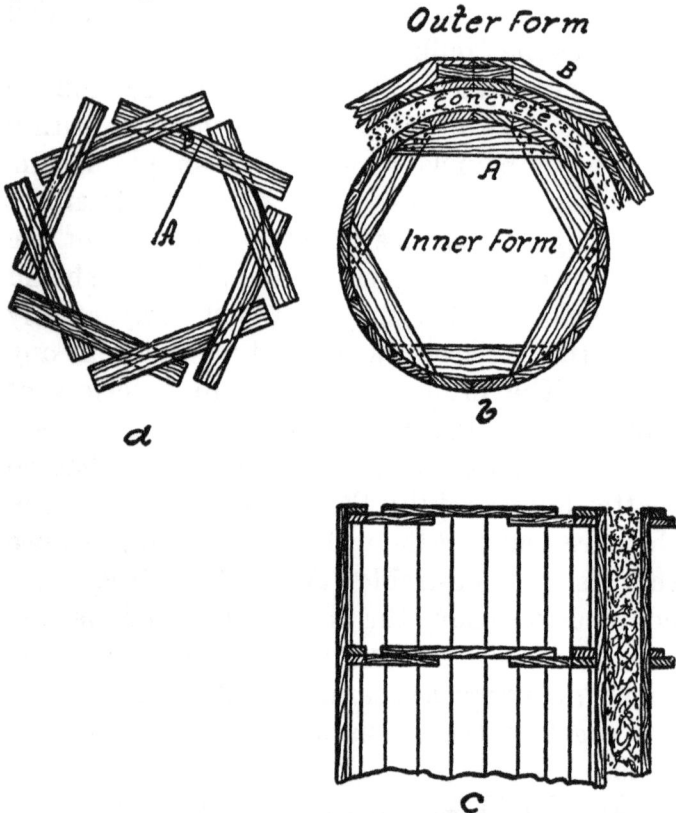

Fig. 2. Method of Constructing Circular Forms.

Circular Forms. In a circular form there are
two sides—the inner and the outer. These may
be used together, as in building a silo; or, as in a
cistern, using the inner form alone; or for a
column, using only the outer form. Both sides

of the form are made in the same way; but the inner and outer sides cannot be made to the same pattern, as the thickness of the walls comes between the parts, making the radius of each side different.

The simplest way to make a circular form is to draw a circle of the size of the form desired, and lay boards around the circumference of the circle as shown in Fig. 2 at **a**. These boards should be lightly tacked together in place, and,

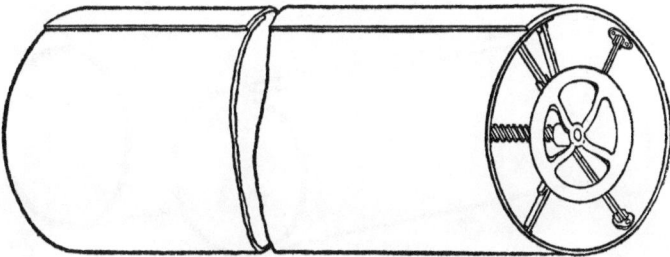

Fig. 3.   Miracle Collapsible Form.

using the same measure, mark a circle upon them. They should then be knocked apart, and sawed out along the lines marked, the pieces being fastened securely together, as shown in diagram. After making two or more forms, place them at equal distances apart, and put on the side boards in the manner shown in the illustration. These boards are called **lagging**.

Culvert Forms

Miracle Moulds. In the many highway improvements that are being made throughout the country, we find a large demand for concrete culverts. The old-fashioned vitrified clay pipe culvert has been found insufficient for this purpose, on account of filling with water from time to time, and freezing and breaking during the winter months.

One of the metal forms used in making culverts is known as the "Miracle." It is of the **collapsible type,** and permits of the use of as

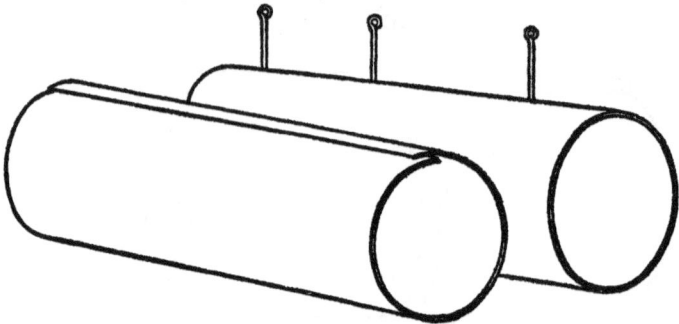

Fig. 4. "Overturf" Collapsible Form.

heavy a wall of concrete as may be required, and also of the use of reinforcing metal, if desired. The trench is dug and the concrete bottom put in place; and the forms are then laid in, the concrete being tamped around them and finished to the desired height and thickness. Then **the forms are collapsed by turning the wheel at the end of the form,** which enables the forms to be withdrawn.

Where one culvert is of insufficient size to carry the water, and it is desired to keep the depth of the same down to a minimum, two or three outlets may be made for the water.

The Overturf Mould. Another mould of the collapsible type is known as the "Overturf." The only lumber needed in culvert construction in connection with this mould is that in the coping moulds. The method of using the Overturf mould is very simple. After the ditch has been dug, the coping forms placed in position, and a bed of concrete laid up to the point that the culvert is to occupy, the mould is laid on this bed of concrete, and set at each end into the forms for it on the coping moulds. The mould is bolted up with bolts extending straight from the top, and all is ready for laying the concrete. After the concrete has set sufficiently, the bolts extending up through it are loosened and screwed up and out, thus releasing the mould, which is self-collapsing, for removal.

Blaw Centering. The "Blaw" collapsible steel centering is used extensively in the construction of sewers, conduits, and other underground work. The centering is in the nature of a steel cylindrical form consisting of plates that are held in position by rods on the interior. When it is desired to remove the forms, these rods are unscrewed, and the form is released.

Systems of Forms for Building Construction

Monolith System. A considerable item in the cost of building concrete walls, as already noted, is the labor and lumber required for the forms. The "Monolith" system of concrete construction is a metal substitute for wooden forms, with separate cores to make a wall hollow.

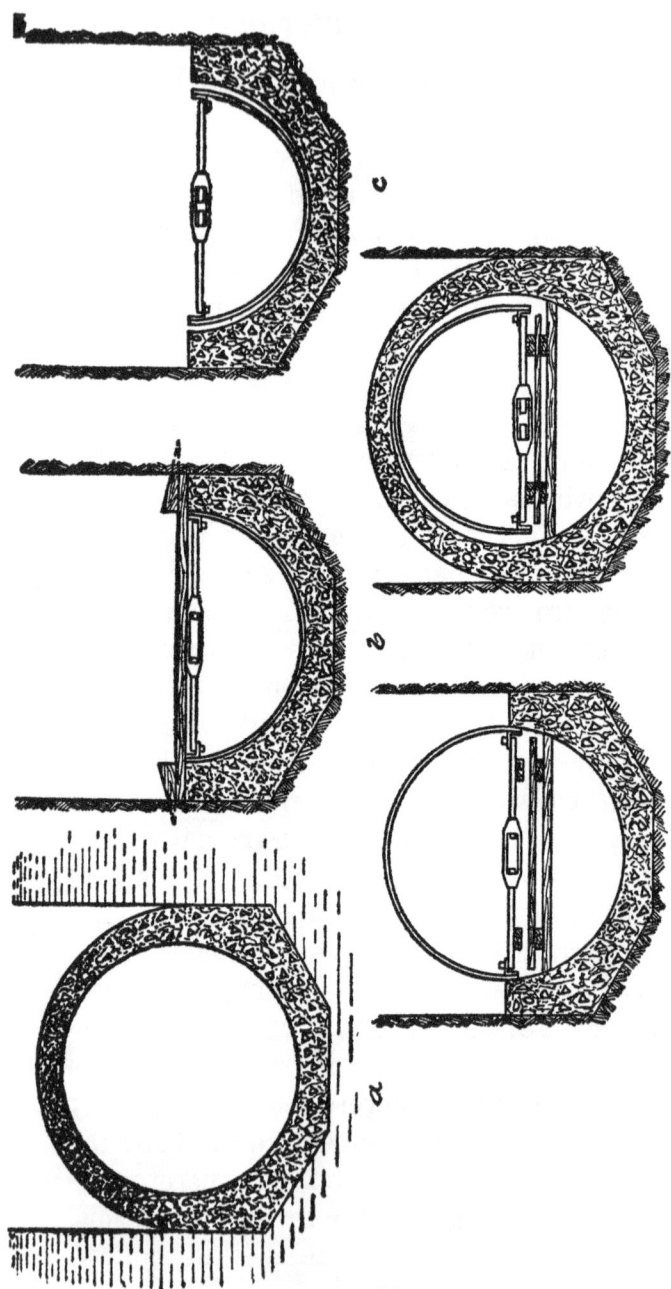

Fig. 5. Construction of Circular Conduit with Collapsible Metal Half-Bound Form.

a—Section of Proposed Conduit. A little concrete is first laid along center of trench; on this, form is placed and held to grade as shown in b. Concrete is then placed, and, after setting, form is collapsed as in c and withdrawn. For crown construction, form is placed as in d, being withdrawn as in e after setting of the concrete.

These can be set together by even unskilled labor to any length or height or width, according to the size and number of forms, and it is claimed that they can be put up in two-thirds of the time required for the erection of wooden forms.

These forms are made of No. 16 gauge galvanized sheet iron, riveted to 1 by 1-inch angle iron. For ordinary use the 24 by 24-inch size has been found most convenient to handle. The

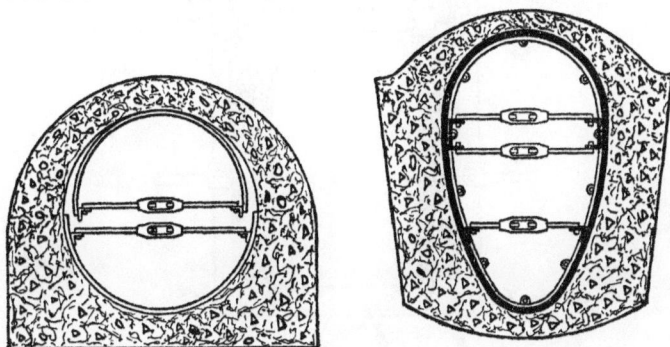

Fig. 6. Blaw Centers Collapsed and in Position.

cores are made of wood, and can be used for any thickness of wall. They are made in several sizes—the cylindrical, 6 in. diameter, for 10-in. and heavier walls; and the 3 by 6-in. for 6-in. to 10-in. walls. They are made in 24-inch lengths, but can be made almost any size. These cores have a cross-bar on top, which has pins at the ends serving to connect the side plates together, and giving the different thicknesses of wall. Tie-rods are required, one for every two feet, but, if desired, may be used only in bottom course, and wire used for the balance. They are furnished

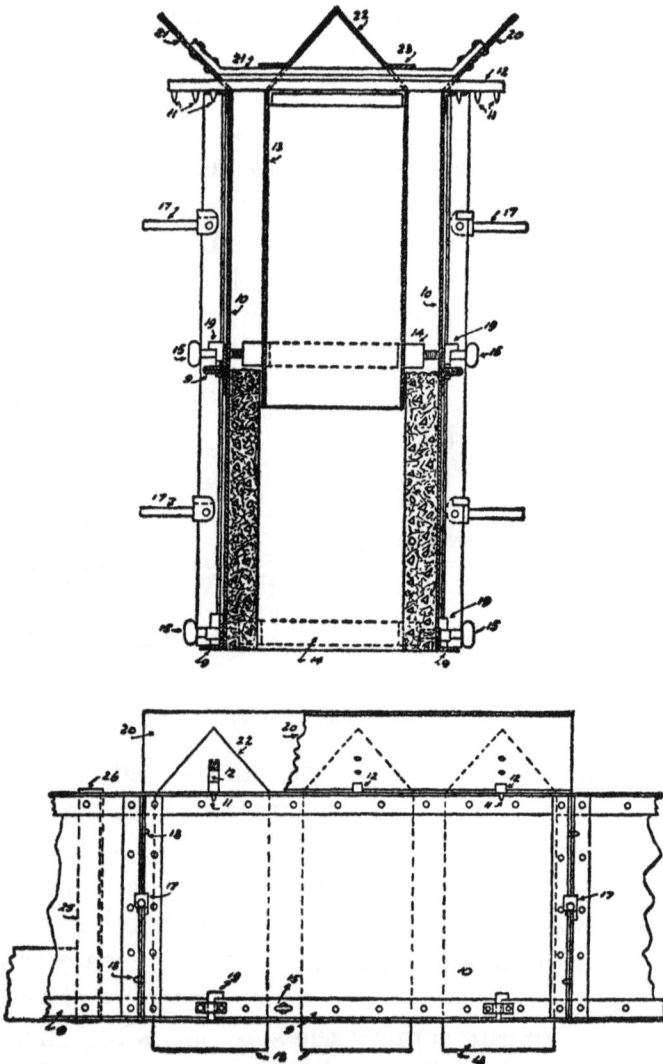

Fig. 7. Details of Monolith Concrete Form.

Fig. 8. Monolith Tie-Rod.

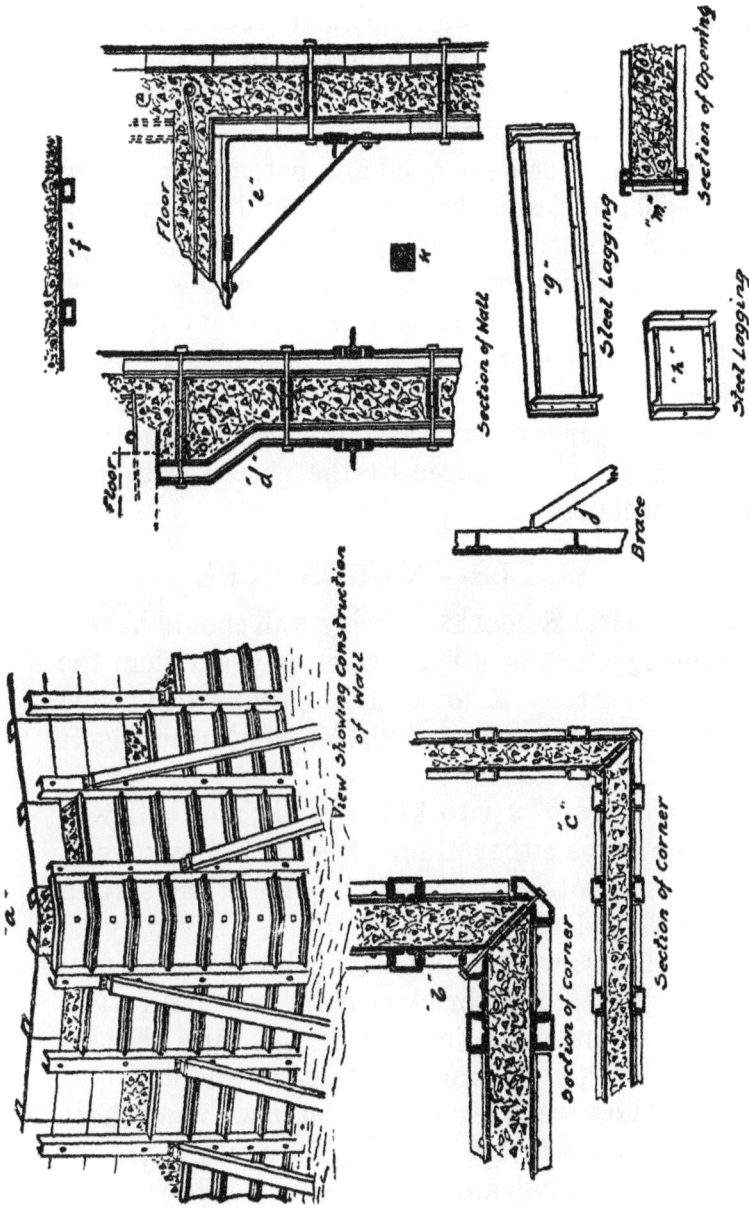

Fig. 9. Details of Construction of Concrete Buildings with Jackson Patented Steel Ribs and Lagging.

for any width of wall. The tie-rods remain in the wall and act as a reinforcement.

The Jackson System. In order to reduce the time and cost of material for concrete structures, there has been developed and patented a system of concrete forms, known as the "Jackson" system, for concrete buildings, as well as for bridge piers, retaining walls, dams, foundations, arches, and reinforced concrete work of every description. The use of this appliance, it is claimed, cuts the time one-third and the cost of forms seventy per cent, and absolute accuracy of lines and surface is assured by the perfect fitting of the steel plates.

WALLS—FOUNDATIONS

General Remarks. Every wall should have a **footing**—that is, a base which is wider than the wall it carries. A foundation must extend below the frost line, and must also extend through soft or yielding soil.

Walls are of two kinds—**solid** and **hollow**—and may be either **plumb**, the same thickness at top as at bottom, or **battered**, wide at the bottom and sloped toward the summit. They may be built in two ways—first, cast in blocks and put in place the same as brick or stone; second, cast in place in one piece (monolithic). Walls must be true, level, and—unless battered—plumb.

Hollow walls are usually built with two faces 3 inches to 4 inches thick, and are either tied together with galvanized iron strips or have piers

of concrete connecting the two faces. These
piers are built at the same time as the faces, and
the whole is practically one wall, with air-
chambers at regular intervals. Walls should be
allowed to season before any superstructure is
built upon them, to prevent their being injured

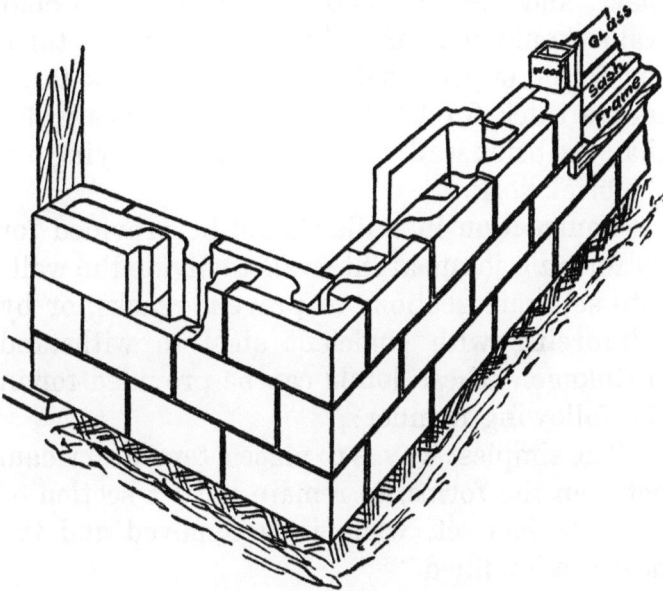

Fig. 10. The "Thomas" Wall.

by workmen. In dry, warm weather, this will
require from six to eight days. Earth should
not be filled in against a concrete wall for three
or four weeks, unless the form farthest from the
earth is kept in place. Where there is no earth
or water pressure against the wall, 24 hours, or
until the concrete will withstand the pressure of
the thumb, is sufficient length of time to keep

forms in place. For this reason it is well to have two or more forms, where movable forms are used. In building forms, they should be assembled, as far as possible, with bolts, so that they may be used again.

Window- or door-frames should be put in place, and the wall built around them. Cellar walls should be from 10 inches to 12 inches thick for frame superstructure; and 14 inches to 24 inches thick for brick, or about 2 inches wider than the brick wall, for convenience in laying out the brickwork.

Contraction in walls should be provided for by forming joints at intervals to divide the walls into separate sections to prevent cracks, or by reinforcing with sufficient steel to withstand shrinkage. These **joints** can be provided for in the following manner:

The simplest way is to place a temporary dam between the forms, to remain until a section of concrete has set, when it is removed and the next section filled.

Another way of forming a joint is to insert two or more thicknesses of tarred paper between sections of the wall.

House Foundations. As a general rule, the base of a wall of any kind should be at least fifty per cent wider than the wall itself. For instance, a 12-inch wall should have a base of 18 inches, which will ordinarily give sufficient thickness to prevent settling. It should be remembered in building the base of a wall, that the entire weight of the building and all its contents is to rest thereon. The stronger the base, the more lasting the building. Another important factor is the necessity of starting a wall below the frost line and below soft or yielding soil.

Excavating. In building the foundation for a house, the first step is the excavation to the desired depth of the cellar or basement. Around the edge, if the house is to be of moderate size, dig a trench eighteen inches wide and six inches deep, and here build forms for a wall of the thickness desired. The dimensions of the trench and footing will, of course, depend upon the thickness of the wall and the weight of the superstructure.

Proportions. The concrete for the wall should be one part cement, two and one-half parts sand, and five parts gravel or broken stone. It should be rammed carefully, and the concrete at the bottom should be allowed to flow out and fill the trench to the desired height. The concrete should be allowed to set hard before removing the forms. In clay soil the outside of the foundation wall should have a good coat of cement mortar. If earth is filled in against the back of the wall, the face forms should be left for three or four weeks, but the superstructure may be begun in about a week, contingent of course upon weather conditions.

Partition Walls. Partition walls are constructed in the same manner as outside walls, but need not be more than 8 inches thick. If re-

inforced with ¼-inch rods spaced 18 inches apart
horizontally and vertically, 4 or 6 inches will be
thick enough for a wall above ground. In wet
ground, to prevent moisture from soaking
through, it is well to give the back of the wall a
coat of one part Portland cement and one part
sand, one-quarter inch thick, applied with trowel
and wooden float, after picking the wall well
with a stone axe and wetting thoroughly. There

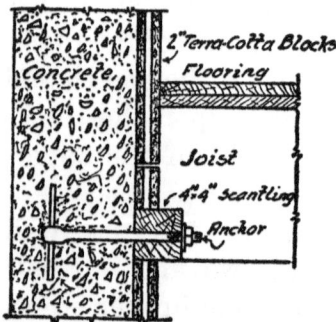

Fig. 11. Section of Wall, Showing Method of Support-
ing Joists.

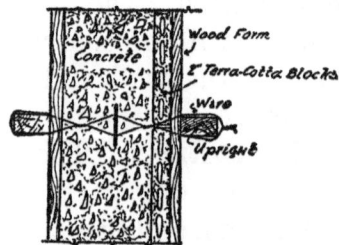

Fig. 12. Wall Lined with Porous Terra-Cotta.

is little danger, however, of moisture passing
through a well-laid wall, if a blind drain of coarse
gravel is laid just back of the foundation, to
carry off the water and prevent its rising back
of the wall. In gravel or sand, the blind drain is
unnecessary.

Concrete Blocks for Foundation Walls. For
basements less than twelve feet in height, a wall
eight inches thick is sufficient for supporting a
two-story frame building. The size of the blocks
is a matter of taste, but no block should be longer

CONCRETING SURFACE OF A DIKE ERECTED FOR PROTECTION OF FARM LANDS FROM OVERFLOW OF THE OHIO RIVER NEAR CINCINNATI, OHIO.

than four times its height, or shorter than one
and one-half times its height. The best builders
favor blocks that are two times as long as they
are high. Crushed stone that passes a three-
quarter-inch screen is best for strength, although
it is rough in appearance. The best proportion
of ingredients is cement one part; sand, three
parts; and crushed stone, one and one-half parts.

Comparative Cost of Walls. This will vary
with the locality and with labor conditions. The
comparative cost of a wall built of brick, concrete
blocks, or monolithic concrete moulded in forms,
may be given approximately as follows: Brick
at $16.00 a thousand laid in the wall. Concrete
blocks will cost $13.80 for the same space; while
monolithic concrete should cost about $11.00 to
fill the same space.

Finishes for Concrete Surfaces. A pleasing
and consistent surface finish generally has but
little to do with the strength of a concrete structure, but it is not inconsistent with maximum
strength in any structure.

Next to form or design, the character of the
surface has most effect on the appearance of
concrete, whether in a building, arch, wall, or
abutment; in fact, when the view is had at very
close range, or in such structures as retaining
walls or pavements, the surface finish may take
precedence over proportion.

We shall describe some methods used in
trying to obtain satisfactory surfaces in the

various classes of concrete work done in the South Park system of the city of Chicago.

The imperfections in the exposed surfaces of concrete are due mainly to well-known causes which may be summed up as follows:

1.   Imperfectly made forms.
2.   Badly mixed concrete.
3.   Carelessly laid concrete.
4.   Efflorescence and discoloration of the surface after the forms are removed.

Forms with a perfectly smooth and even surface are difficult and expensive to secure. Made of wood, as they usually are, it is not practicable to secure boards of exact thickness; joints cannot be made perfectly close; the omission of a nail here and there allows warping; and the result is an unsightly blemish when least wanted.

Badly mixed concrete gives us irregularly colored, pitted, and honeycombed surfaces, with here a patch of smooth mortar and there a patch of broken stone exposed without sufficient mortar. Careless handling and placing will produce the same defects.

But, granting we have the best of labor, that all reasonable expense and care is had in making up forms, in mixing, handling, and placing the concrete, that it is well spaded, grouted, or the forms plastered on the surface, still the results may be unsatisfactory. All these efforts tend to produce a smoothly mortared surface; and the smoother the surface, the more glaring become minor defects. The finer lines of closely-made joints in the forms become prominent, the grain of the wood itself is reproduced in the mortar surface, hair-cracks are liable to form, and, worst of all, efflorescence and discoloration are pretty sure to appear.

It is of doubtful efficiency to line the forms with sheet metal or oilcloth. Imperfections still appear.

Two methods suggest themselves as likely to overcome the defects alluded to above:

(1) Treating the surface in some manner after the forms are removed to correct the defects;

(2) Using for surface finish a mixture which will not take the imprint of the forms and which will minimize rather than exaggerate every imperfection in the latter, and which will not effloresce.

Methods of treating the surface by **bush-hammering, tooling,** and **scrubbing** with wire brushes and water, have been described in various published articles, all of which have for their object the removal of the outer skin of mortar in which the various imperfections exist. But the method most used in the South Park work is the **acid treatment.**

This method of finishing consists in washing the surface with an acid preparation to remove the cement and expose the particles of stone and sand, then with an alkaline solution to remove all free acid, and finally giving it a thorough

cleansing with water. The operation is simple
and always effective. It can be done at any time
after the forms are removed—immediately or
within a month or more. It requires no skilled
labor—only judgment as to how far the acid or
etching process should be carried. It has been
applied with equal success to troweled surfaces,
like pavements; to moulded forms, such as steps,
balusters, coping, flower vases, etc.; and to concrete placed in forms in the usual way. It, of
course, means that in the concrete facing, only
such material shall be used as will not be affected
by acid—such as sand or crushed granite. It
excludes limestone.

The treated surface can be made any desired
color by selection of colored aggregates or by
the addition of mineral pigments. The colors
obtained by selection of colored stone are perhaps the most agreeable, and doubtless the most
durable.

There have been moulded in the South Park
shops blocks for buildings, columns, architectural mouldings, and ornaments with both red
and black crushed granite, all treated with the
acid to bring out the natural colors of the stone.
There has been a large quantity of concrete
pavement laid with torpedo sand surface colored
a buff sandstone color with a small quantity of
yellow ocher and mineral red and treated with
acid. The buff color imparted to the surface is
a welcome relief from the glare of the ordinary
whitish grey concrete pavement in the sunshine;

and the etching of the surface adds to the softness of the color, at the same time preventing any slipperiness. The same buff color has been used to a large extent in steps, bases of lampposts, and other moulded articles to be placed on or near the ground. With sand as the aggregate, thousands of pieces have been moulded for coping, balustrades, concrete seats, drinking fountains, pedestals, etc., which, when treated with the acid, appear like fine-grained, almost white sandstone.

Where there are projections or marks left by the moulds or forms, they are tooled or rubbed down before treatment; and where it is necessary to plaster up rough places or cavities in the surface, it may be done after treatment, and cannot be detected.

These various classes of work have in all cases proved satisfactory.

The second method of preventing or minimizing surface defect has also been tried in the South Park work with quite a measure of success.

During the years 1904, 1905, and 1906, groups of concrete buildings were erected in nine different parks, costing with their accessories, from $65,000 to $150,000 for each group. These buildings are all monolithic structures, with occasional expansion-joints. the exposed surfaces of walls being of concrete composed of one part cement, three parts fine limestone screenings, and three parts crushed limestone, known as the one-fourth-inch size. This was thoroughly mixed quite dry, so that no mortar would flush to the surface, and well rammed in wooden forms made in the usual manner. The result was an evenly grained, finely-honeycombed surface, of a pleasing soft grey color, which grows darker with time and blends admirably with the park landscape. In placing, the concrete was not spaded next the forms; it was too dry to cause any flushing of mortar; so there is no smooth mortar surface, the imprint of joints between the boards is hardly noticed, and the grain of the wood is not seen at all. There is no efflorescence apparent on the surface anywhere, and cannot be on account of the dryness of the mixture and the porosity of the surface. The buildings are used as gymnasiums, assembly halls, reading and refreshment rooms; and, as a rule, the same grey concrete finish is given the interior walls as the exterior. In some cases a little color has been applied on the interior walls, and the walls of shower and bathrooms have been waterproofed with plaster. The porosity of the surface makes it well adapted to receive and hold plaster.

This sort of surface is not capable of treatment with acid as is a smooth mortared surface; nor is such treatment desirable. Consequently the only color obtainable is the natural color of the cement-covered stone, which is softer and far more agreeable than the grey of the usual mortar-finished surface. It is not suited for the surface of a pavement, and is not impervious to water. Although it is evident that water enters the poures to a considerable extent, there is no evidence of injury from the frost during the winter.

The same finish has been used for retaining walls, arch bridges, fence-posts, walls enclosing surface yards, etc. In the buildings the thin walls were made entirely of this mixture, while in the heavier structures it has been used only as a facing. Two reinforced arches of 60 feet span were faced with this mixture, but the steel was imbedded in a wetter, more impervious concrete. The same dry mixture can be used for moulded stones when the mould is open enough to permit tamping, and of course it is eminently suited to block machines.

With the finely crushed stone, a sound, smooth surface was obtained (when the sides of the boxes were removed) where it was manifestly impossible to plaster or grout the surface, and where spading a mixture of coarse stone simply washed the cement away from the surface stones.

Stucco. Stucco-work is cement plastering, and, in one form or another, has been in use for ages. It is durable, artistic, and impervious to weather. For veneering new buildings, or protecting old structures, and wherever the cost of solid concrete is prohibitive, Portland cement stucco has qualities that very highly commend it.

As a rule, two coats are used—the first, a scratch coat composed of 5 parts Portland cement, 12 parts clean, coarse sand, and 3 parts lime, with a small quantity of hair; the second, a finishing coat composed of 1 part Portland cement, 3 parts clean, coarse sand, and 1 part slaked lime paste. Should only one coat be desired, the finishing coat is used. Some masons prefer a mortar in which no lime is used, but this requires more time to apply.

In applying stucco to brick or stone structures, clean the surface of the wall, and, after thoroughly wetting, plaster 1½ inches thick. For a finish, either smooth with a wooden float, or rough by rubbing with burlap.

In using stucco on a frame structure, first cover the surface with two thicknesses of roofing paper. Next put on furring strips about one foot apart; and on these fasten wire lathing. (There are several kinds, any of which are good.) Apply the scratch coat ½ inch thick, and press it partly through the openings in the lath, roughing the surface with a stick or trowel. Allow this to set well, and apply the finishing coat ½ inch to 1 inch thick. This coat can be put on and smoothed with a wooden float, or it can be thrown on with a trowel or large stiff-fibered brush, if a spatter-dash finish is desired. A pebble-dash finish may be obtained with a final coat of one part Portland cement, three parts coarse sand and pebbles not over ¼ inch in diameter, thrown on with a trowel.

Cutting with Hammer. Eight or nine cuts to the inch with a hammer, also makes a good finish. The surface also may be "picked" with a brush hammer. To do this and produce anything like good results, the forms must be removed before the concrete has had its final set, and the surface gone over with a wire or stiff bamboo brush. Brushes of this kind will remove the skin or scale of cement, and leave the stones of the aggregates exposed in their natural color. For the producing of good effects in this finish, aggregates of varying color should be used.

Securing an Artistic Surface. Albert Moyer gives this suggestion for securing an artistic surface:

"In using concrete for country residences, I wish the reader to eliminate from his mind all thought of concrete such as he sees about him in retaining walls, bridge abutments, and other work where concrete has been employed, but to try to picture a concrete made of selected materials, the moulds or forms taken off as soon as possible while the concrete is yet green, the surface scrubbed with a scrubbing brush, or, if the concrete is too stiff, with a wire brush, water being sprayed on with a hose, thus removing all the mortar which has come to the surface, and exposing the larger pieces of aggregates—in fact, throwing them slightly in relief, giving a rough surface of accidentally distributed colored stones.

"As the walls are erected in different courses,

the lower courses are from necessity stained by surplus water running down from the upper forms. This is very readily removed by washing off the walls after the house is completed with commercial muriatic acid, 4 to 6 parts water, which further brightens up the different particles of stone and removes any cement stain that may be on the outside surface of the stone or on the mortar which bonds the stones together.''

Painting the Surface. In some localities a popular method of finish, as well as the cheapest, is to paint the surface with cement mortar. There is danger, however, in this method, of marring the effect by drippings if the work is done carelessly.

Use of Colored Aggregates. Pleasing effects are produced by the use of aggregates of uniform color, like crushed red stone or blue stone. The red stone gives a pinkish shade to the wall in conjunction with the cement. If a brush is used before the concrete is dry or set, these colored aggregates can be made to stand out in relief.

Metal Lath for Building

Concrete steel exterior walls, such as are made ordinarily in the construction of residences, can be made in various ways. Expanded metal of three-quarter-inch mesh is often used, being fastened to studding with sixteen-inch centers. This is considered superior to any other method, as the mesh is sufficient to permit the cement to pass through and cover the other side of the lath, thus protecting it.   The ordinary sheet-metal lath with protruding cups and opening, cannot be covered on both sides.   This variety is, therefore, not recommended for exterior walls, as dampness will soon rust the metal sheet.   Another method satisfactory and effective is to cover the outside of the building with sheathing, and place the plaster lath vertically about one foot apart.   On this, poultry netting of three-quarter-inch mesh is placed, and the plaster is applied.   This will require less material than any other method.

ASBESTOS BUILDING LUMBER

The fire-resisting, non-conducting qualities of the mineral **asbestos** have long been well known.   This natural rock of fibrous texture is manufactured into a great variety of commercial products adapted to various uses where insulation is required.   One of the latest applications of this useful agent is as **building lumber**, in which form the mineral fiber is manufactured into large sheets which can be installed in building construction in the same way as ordinary lumber for floors, partitions, etc.

The use of large sheets of asbestos building lumber as a fire stop in partitions, between the rooms of summer hotels or in office buildings where space is of great value, is often exceedingly desirable, as by the use of small I-beams, or angle-iron, or studding, or even ordinary common board partitions between rooms, faced upon both sides with ¼-inch asbestos building lumber, a thin, strong, fireproof separation is made, that is very saving of space. Partitions of light wooden studding faced with asbestos building lumber up on both sides, are likewise of most excellent construction as a fire stop, particularly when 85 per cent magnesia blocks are placed between the studs, making the partition not only a non-conductor of sound, but rendering it almost impossible for any ordinary fire to pass beyond such a barrier.

Where it is desirable not only to cut down insurance rates, but, for one's own personal safety, to confine the fire that may unfortunately occur in any dwelling, to the floor upon which it originates, these objects may be accomplished by the use of asbestos building lumber in large sheets, upon the rough floor, when the finished floor may then be laid immediately upon these sheets, the whole being securely nailed in the usual manner. This will confine any ordinary fire to the floor in which it originates, and preserve the lives and treasures of the occupants. Of course, in house construction, the sheets of asbestos building lumber must run underneath the studding, to prevent the sparks and flame carried by the air-draught from coursing through the partitions.

As this form of material is too hard for rats

and mice to gnaw through, it affords an effective barrier to the ravages of these pests.

ROOFS, CHIMNEYS, ETC.

Concrete Roofs. It will be impossible to build a concrete roof that will not crack, unless the walls are rightly constructed and on a foundation that will not settle. Roofs require special care to render them water-tight. The sinking of the brickwork on new structures is often the cause of cracks in a concrete roof. Of course a crack in a roof of this material is more serious than in any other work, as it means a leaky building. The rough coat should be laid and solidly rammed and compressed. So far as possible, the top coat should be laid in one piece. Roofs exposed to the sun should be kept damp for a number of days after being laid, as they are easily affected by the heat. Expansion may be counteracted by the use of strips around the walls. A skirting six inches high and one inch thick should be keyed to the walls. A good top coating is made with one part Portland cement, one-half part of slaked lime, and one part of fire-brick dust.

Cement shingles of various shapes and designs—in some cases reinforced with fine wire mesh—are beginning to be used extensively, and are giving the best of satisfaction when properly laid. They are applied in the same way as slate, and provision must be carefully made to avoid cracking during the process of laying. In some instances this is done by inserting in the body of the shingle small pieces of asbestos board through which the nails or screws may be driven.

Cement Brick for Chimneys. Concrete blocks and cement brick make desirable chimneys; and, if care is used in the construction, entire satisfaction ought to be attained. Dry concrete being fireproof to the extent of its raw material, it has been found that sand usually endures more heat than cement. It is therefore necessary to select a cement that has been highly burned—no less than 1,200 degrees—which will make a chimney safe. But the chimney may discolor at 800 degrees without injury. For wood and soft coal fires, Portland cement is acceptable. For hard coal and coke, the cement must be selected, and gravel, limestone, and soft sandstone must be omitted. The inside of the chimney should be plastered with mortar made of one part cement to three parts of sand, mixed with strong salt water.

Cement Roofing. The practicability of cement roofing is no longer a question. It was introduced in the United States several years ago, and its use spread very rapidly, so that we now find its application in many States, on nearly every class of building. There are a number of concerns manufacturing concrete roofings of one type or another, nearly all of which turn out a product which is giving universal satisfaction.

The question of obtaining satisfactory roofing for industrial plants has been a perplexing problem for some time. There are a number of roof coverings which give satisfactory service for a time; but, when subjected to the elements, it is only a matter of a short period until they are partially or totally destroyed. The fumes and gases from furnaces and factories, prevalent in manufacturing districts, attack the old-time roofing, and accomplish the work of destruction in a comparatively short time. The natural elements bring the same results in a somewhat longer period.

Cement roofing has proven itself economical, practical, and durable. While the original cost of this kind of roofing is somewhat more than that of slate or other materials, in the end the concrete covering always proves the most economical.

Cement **roofing tile** have proven their adaptability on a large number of buildings of prominent industrial concerns. The tile are reinforced. The weight, which will about equal that of a slate roof on two-inch sheathing, is in the neighborhood of thirteen pounds to the square foot. The tile are about seven-eighths of an inch in thickness, cover a space four feet long by two feet wide, and are attached directly to the purlins. A permanent red color is obtained in these tile, which is claimed to be non-effaceable by the elements, requiring no painting. The tile are self-adjustable and interlocking, which provides in every way for expansion, contraction, and vibration, and at the same time afford a covering unaffected by temperature changes, and one which is fireproof and waterproof. See Plate 1, upper figure.

Every fire test of properly made concrete, whether made by Fire Underwriters or the Government Laboratories, proves this modern building material to be a most effective fire-resistant. This advantage alone, without doubt, has, more than any other factor, placed concrete roof covering in the position it now holds. The permanently waterproof qualities of concrete roofing also give it an advantage over other types of roofing likely to be affected by the heat of the sun and the destructive work of the elements.

The "Diamond" cement shingle is adapted to all manner of roofs, hips and valleys, without the necessity of cutting or fitting. It is made by the "face-down" process. These shingles are nailed to the sheathing with sixpenny wire nails. No expensive copper wires are needed. Where roofing is wired on, it is difficult to use paper or felt; but where nails are used, as with the present type, it is claimed that the use of slater's felt or tarred paper is not only possible but desirable. See Plate 2.

The Grumman cement shingle is a type of the reinforced class, being reinforced with expanded metal lath. The fact that the operator finishes the shingles—as also the floor tile made under the same system—on the top or face, enables him to color his shingles and tile any desired color by mixing cement and coloring, equal

REINFORCED CONCRETE ROOFING TILE.

INTERIOR OF A CONCRETE CATTLE-BARN.

PLATE I—CEMENT CONSTRUCTION.

parts, and sifting on shingle or tile, then troweling off to a finish. See Plate 2.

Another type of the cement shingle is the "Superior." They are of suitable dimensions for any style of building on which shingles, slate, or tile may be used. They do not hang loose on the nails, but are nailed down like wood shingles. The abestos nail-pads being sufficiently pliable to take care of any contraction or expansion, and, by reason of their peculiar construction, rendering impossible capillary attraction of moisture. With their wire reinforcement it is claimed that they are much stronger than any slate and practically everlasting, while at the same time adapting themselves well to ornamentation.

They are laid on the roof 7 inches to the weather, thus requiring 294 to the square. The weight is 600 lbs. per square. The cost to manufacture may be easily estimated, calculating 400 lbs. of sand, 200 lbs. of cement, with the asbestos nail-pads and wire reinforcement, all subject to local market conditions, which, as well as labor, differ in various localities.

A form of reinforced concrete roofing that requires no centering is exemplified in the "Ferro-Lithic" plate. This type of roof plate consists of iron sheeting so corrugated that the openings at once form a bond for concrete on the exterior and plaster underneath. These plates are used also for siding, stairways, floors, etc., and form both permanent centering and reinforcing. They are quickly laid in place, and easily concreted on top and plastered on the under side. See Plate 3.

The vertical portion of the dovetails serves as ample reinforcing, being greatly assisted by the mechanical bond secured by the heavy cross-ribs.

When completed, this roof is claimed to be the lightest type of reinforced concrete roof on the market—a fact of considerable value in the design of roof trusses.

Inflammable roofs are a constant source of fire communication, but this roofing is fire-resistant and protective. It may be laid and used without concreting, during stormy or wintry weather, and concreted at a more favorable season.

Asbestos Shingles. A combination of asbestos and Portland cement has been produced, which, when made up in the form of shingles, furnishes a roof that is absolutely fireproof. The shingles are made under high hydraulic pressure. Simple exposure to the elements causes the cement that has been deposited in the asbestos fiber to crystallize, and the material then becomes better, tougher, and harder as time goes on.

Another good point for these shingles is that they do not need paint, or any attention, as the elements take better care of asbestos shingles than any paint or dressing.

These shingles, owing to their composition, are sufficiently elastic to allow of marked tension due to vibration, expansion, and contraction of surrounding parts, etc., without breaking, cracking, or tearing apart. Their resistance to blows, tensions, or lesions of any sort is said to be surprising. They can be worked easily with ordinary tools. They become very hard in time, especially when exposed to the weather.

They are usually applied by the French, or diagonal, method. Shingles for a finished roof, laid in this manner, weigh about 250 pounds to the square. This lightness, with its accompanying efficiency, renders them peculiarly adaptable to lighter construction.

CEMENT FLOORS AND STEPS

Sarco Mastic. For manufacturing establishments where it is desired to have absolutely waterproof floors, "sarco mastic" is used in connection with "sarco flux" on cement floors. The mastic is broken up into small pieces, and placed in a suitable kettle, together with a certain percentage of the flux, and heated to about 450° F. After the materials have been cooked at the above temperature for from two to four hours—depending upon the size of the batch and the construction of the mastic boiler—a certain percentage of grit should be added. This grit is procured in the vicinity of the work, and should be entirely free from loam and dirt, not larger than No. 4 torpedo or fine limestone or granite screenings. The grit is thoroughly stirred into the mass with suitable stirring rods, and subjected to the same degree of heat that was used to break down the mastic, for a period of one to two hours, depending, as before, on the size of the batch and the construction of the mastic boiler.

After the material has been thoroughly cooked and mixed, it is shoveled from the boilers into oak buckets or into wheelbarrows, care being taken (in case of carrying any distance from the boiler) to keep the material covered to hold the heat.

When applying to the surface to be covered, the material is spread out as much as possible by the carrier, and then distributed evenly over the floor to the proper thickness by the spreader, and rubbed smooth.

Veranda Floors. Concrete floors for verandas, porches, and even balconies are much to be desired. While they cost a little more than wood, they are imperishable. Such floors may be built on the ground the same as walks, or on rough wooden supports. One part of Portland cement, three parts of sharp sand, and three to five parts of gravel, mixed with sufficient water to become plastic, and tamped in position to a thickness of 3 inches, is the rule. The wearing coat is a half-inch thick, and made of one part Portland cement and two parts sand. The top coat should be smoothed with a plasterer's trowel, and allowed to stand for six to eight days, being kept damp.

Jointless Floors. The use of jointless flooring, made from pulverized wood fiber and other material, and laid in a plastic state on a cement foundation, was begun in Germany about ten years ago. This flooring has proved so successful that several other mills manufacturing the same product have been started, and are now running prosperously. The problem has been to make a continuous flooring, which will fit closely at its junctions with the upright walls, and be not only fireproof, but impervious to liquids, dust, and vermin of all kinds. The experts at work on the problem hope to succeed in producing a flooring, too, that will be a poor conductor of heat and sound, easily cleaned, neat, and attractive.

Cellar Floors. Cellar floors may be laid without foundations, except in places where there is danger of frost getting into the ground below the floor. The dirt should be evened off and tamped hard, and the concrete, one part Portland cement, two and one-half parts clean, coarse sand, and five parts broken stone, spread over the surface in one continuous slab 3 inches to 4 inches thick, lightly tamped to bring the water to the surface, and screeded with a straight-edge resting upon scantlings placed about 12 feet apart. The scantlings are then withdrawn, and their places filled with concrete. No finishing coat is needed unless the floor is to have excessive wear. The surface of the concrete, however, should be troweled as soon as it has begun to stiffen. Joints about 12 feet apart should be made if the surface is more than 50 feet long, or if it is to be subjected to extreme temperatures.

Flying Stairs or Steps. In constructing "flying" steps or stairs, first make the steps as in Fig. 13 and allow them to season. The stringers are cast in place, being constructed as inclined beams of sufficient length, and with a projection along the lower inside to support the steps. Place two ¾-inch bars about 1½ inches in from bottom, and one of the same diameter at the top, the three being connected. The reinforcing, however, depends upon the depth and

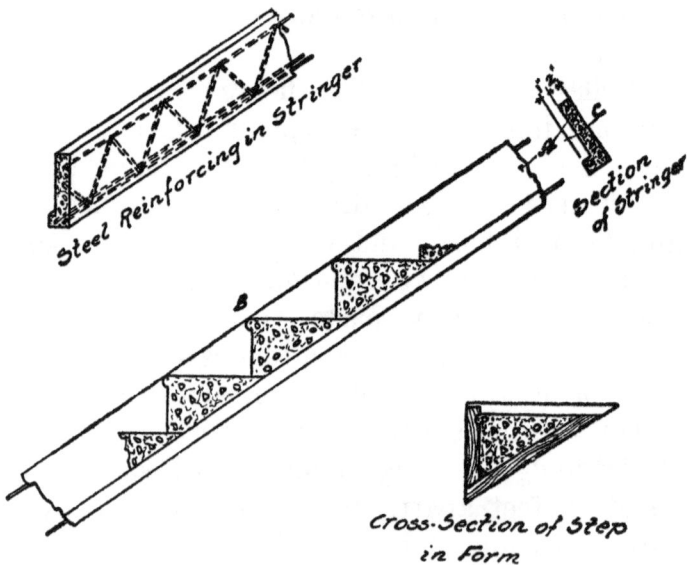

Fig. 13. Construction of Flying Stairs or Steps.

pitch of the stairs, also on the weight they are to carry. Stairs are put in place with the lower edge resting against the upper corner of the stair below, a small notch being left in each stair for this purpose. A finishing coat of one part Portland cement and one part clean, coarse sand is given to the stairs after all are in place and have been picked with a stone ax and wet. This binds the whole flight together.

Steps for Residences. The growing demand for concrete steps for residence entrances has led to the construction of the type of step shown in Fig. 13A. These steps or treads are made at the hollow block plant and kept in stock. The mould, a cross-section of which is shown, is inexpensive and allows the use of rich mix for facing, and wire cloth or expanded metal for reinforcing, being practically a face-down mould with a round corner on the skids for tilting up before removing the pallet.

The tread and riser are adapted to any height of riser, as a study of the section steps shows. Should a narrow tread be desired, it is easy to trim down the pointed edge projection against the riser above.

The cost of these treads is less than solid ones, there being a saving of half the concrete, a saving of labor in applying the richer facing, and only two-fifths the weight to handle.

Only one mould is necessary, but the removable core should consist of several short blocks, so that any length can be met.

The pallets should be of sufficient length for the longest steps required, and the ends of the

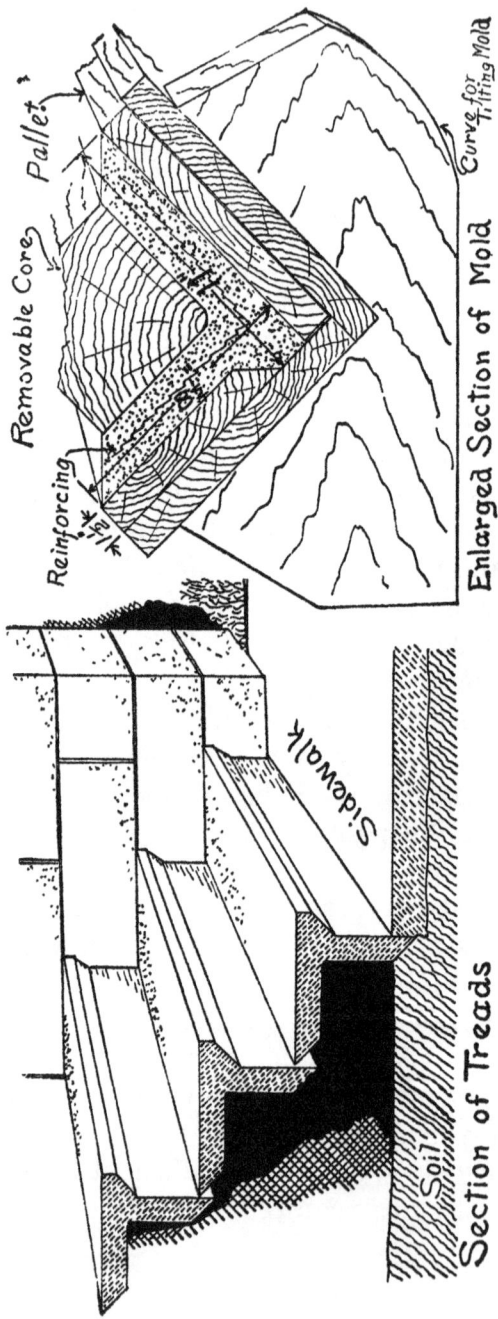

Fig. 13A.   Construction of Concrete Steps for Residence.

Section of Treads

Sidewalk

Soil

Removable Core

Pallet

Reinforcing

Enlarged Section of Mold

Curve for Tilting Mold

mould sliding over the pallet permits short lengths being made on long pallets.

When return ends are desired, a special end board is necessary, which, with an extension on the end of core, admits making a full end of step, thus making the supporting wall at end of steps unnecessary.

When making cement walks, it is surprising to learn how much time these treads save over the old method of using forms; besides, for adjusting a number of steps to meet a given height, these treads have no rival.

Only two-inch surfaced lumber should be used for making the mould, and the skids should not be more than three feet apart. The removable core need not be solid as shown; but this is preferable, for, by tamping this core after the mould is filled, the concrete becomes very compact.

Round rods can be used for reinforcing, but wire cloth or expanded metal lathing is preferable, being more binding and nearly always cheaper.

The supports for these steps are the same as for stone—namely, supporting walls under each end, also a center wall under center for lengths of six feet and over.

Concrete Steps in Damp Places. Concrete may be advantageously used in the construction of steps, particularly in damp places, such as areaways and cellars of houses; and in the open, where the ground is terraced, concrete steps and walks can be made exceedingly attractive. Where the ground is firm, it may be cut away as nearly as possible in the form of steps, with each step left two or three inches below its finished level. The steps are formed, beginning at the top, by depositing the concrete behind vertical boards so placed as to give the necessary thickness to the risers and projecting high enough to serve as a guide in leveling off the tread. Such steps may be reinforced where greater strength is desired, or where there is danger of cracking due to settlement of the ground.

Where the nature of the ground will not admit of its being cut away in the forms of steps, the risers are moulded between two vertical forms. The front one may be a smooth board; but the other should be a piece of thin sheet metal, which is more easily removed after the earth has been tamped in behind it. A simple method of reinforcing steps is to place a half-inch steel rod in each corner, and thread these with quarter-inch rods bent to the shape of the steps, as desired, the latter being placed about 2 feet apart. For this class of work, a rich Portland cement concrete is recommended, with the use of stone or gravel under one-half inch in size. Steps may be given a half-inch wearing surface of cement mortar mixed in the proportion of 1 part cement to 2 parts sand. This system, as well as many others, is well adapted for stairways in houses.

Steps on Terraced Grounds. Excavate on

the slope, allowing for four inches of sub-foundation and four inches of concrete. Put in subfoundation of cinders or broken stone or bricks, providing a drain at the lower end to carry off any water that may accumulate. Concrete work often is ruined by water freezing under it and expanding. Place a plank along each side of the proposed steps, providing one wide enough to take in the rise of each step. The planks should be well braced their entire length to prevent any bulging. Lay a strip of woven wire fabric or other reinforcing mesh, of a width nearly corresponding to the width of the steps, and the full length of the steps on the slope.

The next operation is to spread on the wire a layer of concrete about three inches thick consisting of one part Portland cement, three parts clean, coarse sand, and six parts gravel or broken stone. Sufficient water should be used to make a mixture that will work through the wire cloth and completely surround it. Tamp well, and permit the concrete to stand for twenty-four hours. Starting at the top, place the boards between the planks to form the risers of the steps. The inner top edge of each board should be grooved in a circular form to form the bottom of the nosing of the tread. Each board should be fastened securely to the planks. Just before the next mixture of concrete is applied, the base of the concrete laid before should be wetted. The forms should be filled with a mixture consisting of cement mortar—one part cement, and two parts of clean, sharp sand. Trowel the top, and round at the edge to conform with the groove in the riser board.

Porch Steps. Build two 8-inch walls to a depth below the frost line, and make sure that the upper surface conforms to the desired pitch of the steps, but three inches below the point where the inner edge of the tread meets the rise. Between the walls, build a platform out of 2 by 4 stuff, well braced and conforming to the slope of the walls. Over this, and over the top edge of the walls, put a three-inch layer of concrete reinforced every foot by $\frac{1}{4}$-inch iron bars running from top to bottom. Build up the form on the outside of the walls, and proceed in the same manner as for terrace steps. Should the steps be more than 6 feet wide, a wall similar to the two side walls should be built in the center, to assure sufficient strength. The forms should not be removed from under the steps for 28 days. If more than three or four steps are required, the iron rods must be of larger size and nearer together.

Platform Steps. The best proportion to use is one part cement, and four to five parts sand. Blocks thus made will resist wear well. For the wearing surface, one part cement and two parts sand will be better; but the wearing coat need be only half an inch thick, provided both coats are mixed and used at the same time, allowing the cement in both mixtures to unite. Wood moulds made of surfaced lumber coated with liquid shellac, oil, or paraffin, will not adhere to the cement. The proportion of cement and sand is necessary to secure uniform and sound work as well as color. Plaster for the outside work should be made of cement, as lime disintegrates; besides, fresh lime requires much more water than cement, hence they never should be mixed together. Cement always contains sufficient lime that has been properly treated, so that no more lime is required unless a cheaper product is wanted.

Cast Steps. Steps made in this manner should be given plenty of time to season before they are used or put in position. The making of cast steps is similar in a general way to the making of cement blocks, in that a form is used. This form should be of plank, and the inner surface should be smoothly dressed. One side of the top edge should be grooved to provide for the projection of the tread over the riser. The concrete should consist of one part cement, three parts sand, and six parts broken $\frac{1}{2}$-inch stone. Fill the box or form with the mixture to within one inch of the top, leaving that space for the finishing coat, which should be put on immediately. Dress the top with 1 part cement and $1\frac{1}{2}$ parts sand, rounding the edge to conform with the groove of the projection. After the cement is set and hardened, the form may be removed, and used for another step.

Side Walls for Steps. In building the side walls for steps of this character, the foundation

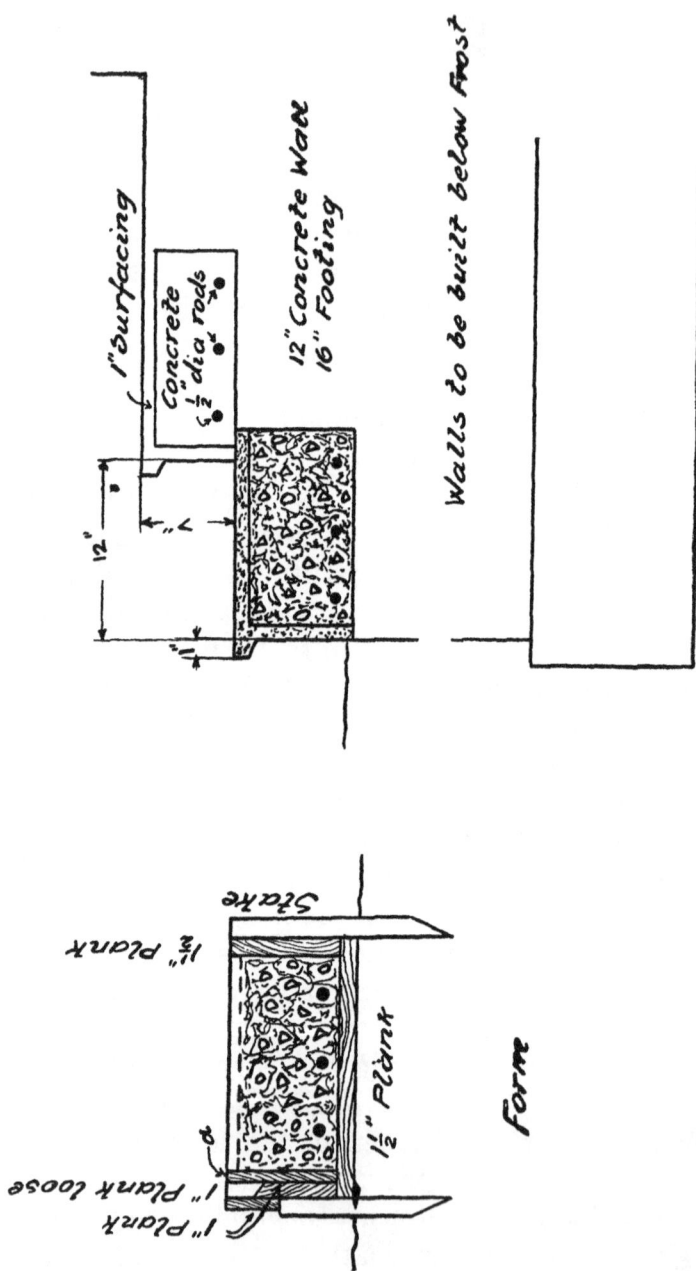

Fig. 14. Details for Making Cast Concrete Steps.

1" Surfacing

Concrete ½" dia rods

12" Concrete Wall
16" Footing

Walls to be built below Frost

12"

1"

Form

1½" Plank

1" Plank loose

1" Plank

1½" Plank

Stake

62

should be twelve inches wide, and should start below the frost line. On this foundation, and at equal distances from each edge, erect eight-inch walls, stepped off to conform to the width and depth of the cast steps. Place the cast steps on the walls thus made, covering the joints and finishing neatly with cement mortar. If the steps are more than three feet long, they should be reinforced with half-inch iron rods placed in the center of the steps, about three inches apart.

Making the Slabs. The formula for concrete used in this work is 1 part Portland cement mixed dry, to 2½ parts of crushed slag and

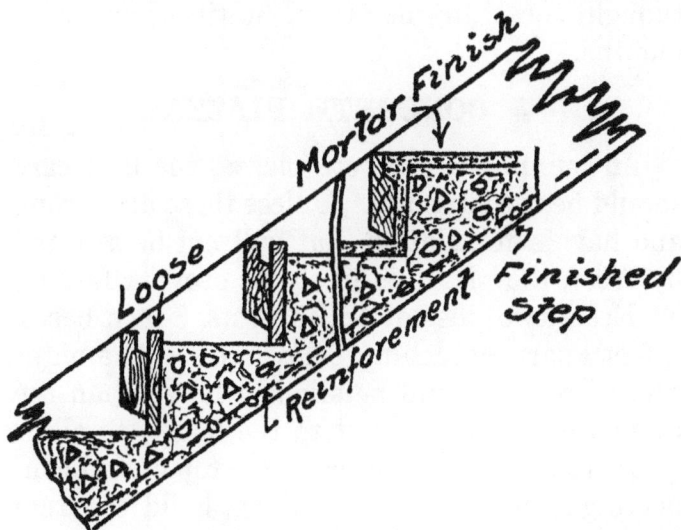

Fig. 15. Details of Concrete Step Construction.

granite equally. The granite and slag should be of a size that will pass through a ⅓-inch sieve. The method of mixing is mechanical. The parts are placed in a horizontal cylinder and thoroughly mixed. The mould is placed on a shaking machine after being previously oiled; and after the machine, which gives to the mould a rapid vertical motion, is started, the concrete is laid on the mould in small quantities, and it is smoothed off with a trowel. The moulds are removed, and permitted to stand for three days. Then the slabs are taken out, and stand five days more. A silicate bath is then used for immersing the slabs, and they are allowed to remain in this a week. After this process, the slabs are taken out, and are dried in the open air. It is not thought best to use them until after three months.

A CONCRETE PIAZZA

In building a concrete piazza, the first care should be the supports. Unless these are strong and have a foundation that will not be affected by frost, the piazza is likely to prove a failure.

Erect two lines of 4-inch posts, 8-inch bases, 8 feet apart, extending below frost. The outer line of posts should be slightly lower than the inner line, which is next to the house, to allow water to flow off the piazza. On top of and connecting these in both directions, build concrete cross-beams and stringers 4 inches by 8 inches. Posts should be reinforced with a ⅜-inch steel bar, and beams with two ⅜-inch bars placed one inch above the bottom.

After the concrete has set hard, erect forms

DIAMOND-SHAPED CONCRETE SHINGLES.

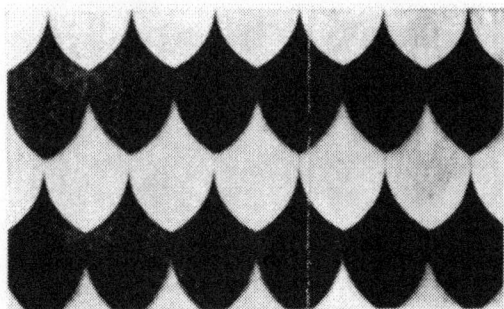

GRUMMAN CONCRETE SHINGLES.

PLATE 2—CEMENT CONSTRUCTION.

and build a solid slab of concrete over the entire framework, allowing it to project slightly over the outer edge. This slab should be reinforced with a woven wire fabric or expanded metal or with steel rods.

If preferred, the forms for the walls and floor may be built at the same time; but in any case, the walls should be allowed to set before the slab is poured.

A finished surface can be obtained by plastering the surface one-half inch thick with mortar— one part Portland cement, and one part clean, coarse sand—before the concrete has set, and troweling it hard as the mortar begins to stiffen.

Lattice Construction. In building a lattice, the fact that there are two thicknesses of concrete—namely, the thickness of the panel or border and the thickness of the lattice itself—should be borne in mind.

A CEMENT COPING

A coping made of cement should be about 4 inches thick, to look well. Take two 6-inch boards and nail on a 1 by 2 strip on the bottom, to form the projection of the stone. The strips can be put on to show any thickness of stone desired, but 4 inches thick is a good thickness for 8- and 12-inch brick walls. Under the side boards, put 2 by 2 strips perpendicular to the ground, or any strips sufficient to hold the form at its proper height. Next, brace the form as shown, in order to keep the form from spreading when the cement is put in. Short forms will not need bracing, but any form six to eight feet long should have at least one brace each side to hold it firmly to the wall. Sometimes, when it is difficult to get a brace in, a board notched just right, to fit over the top of the mould as shown, will answer; but this is not as good as the braces, for it has a tendency to make the mould pull away from the wall at the bottom. After the mould is put in place, take some thick cement mortar, and go over the bottom, stopping all cracks in the mould, for there will be some places where the mould will not fit the brick wall closely enough. It should be made tight with thick cement mortar, well troweled, so that water will not drip through and deface the brickwork. The mould can be made water-tight in a few minutes, and very little mortar will be required for this.

As soon as the mould is ready, mix the sand and cement in proportions of 1 part cement to 2 parts or 3 parts sand, according to how good you want the stone. One to three will make a good job, but 1 to 2 will be extra good. Mix the sand and cement dry first, and mix it thoroughly.

Good cement work depends much on good mixing, and this point should never be neglected. After mixing the cement, put it in the mould, and with a trowel work it down along the sides of the mould until it is solid and without air-holes; and fill the mould as you go, and rather quickly. After the mould is full, level if off on top, and trowel it to an even surface. Let it set a little, and trowel again. After it has set sufficiently to stand without running, then, with a trowel, you can clip the top corners, and trowel them down smooth, thus avoiding the sharp edges on top of the coping which are always getting chipped off more or less, giving the stone a ragged edge. The corner, cut out a little, obviates this; it takes but very little time, and insures a job that will look better and last longer.

The sand and cement should be mixed to a good stiff mortar, not so that it is sloppy. Use the trowel to force the mortar down in the mould and make it fill. Troweling brings the water to the surface; and if mortar is too wet, it floats the cement off the top of the stone, showing sand patches. Watch the troweling closely; and if the mortar is too wet to trowel well, wait a little. A little experience will be your best teacher, and you can soon learn to turn out first-class jobs in cement work.

CHIMNEY CAPS

Chimney caps of concrete are rapidly supplanting those of stone, brick, or iron, as they are not only cheaper and more durable, but protect the top of a chimney better.

Make a bottomless box the size of the required cap, and one or more small bottomless boxes to correspond to the flue or flues of the chimney, and one-half inch higher, so that the surface of the concrete can be sloped to allow water to flow off, and set in place. The thickness is usually about four inches, but this can be varied to suit convenience. Plaster the inside surface of the large mould with one-half inch of stiff mortar, and then immediately fill form one-half full with one part Portland cement, three

Fig. 16.   Form for Making Chimney-Cap.

parts clean, coarse sand, and six parts broken stone, and put in reinforcing, either woven wire, expanded metal, or $\frac{1}{4}$-inch rods, complete, and tamp until water puddles on top. When partly set, trowel smooth.

If it is desired to build the cap in place, the following plan should be adhered to: Place small rods across the chimney between the flues. On these, build platform of tongued-and-grooved boards planed on upper side, and driven snugly together, but not nailed. On this platform, place the forms previously described, and fill with reinforced concrete. After the concrete has set (at least a week is needed) remove platform by raising each side of chimney cap alternately and knocking platform apart. Remove outer and inner forms. Raise one end of slab; cover all accessible surface of top of chimney with mortar; lower cap on bed thus formed; and remove rods under end. Repeat process at opposite end.

A CONCRETE FIREPLACE

A special formula for the blocks used in the construction of a fireplace is necessary to guard against discolorations from the heat. Blocks made in the following proportions will prove satisfactory for the fireplace and chimney construction: One part Portland cement burned at 900 degrees or more, one part fine sand, three parts coarse sand. Blocks must be well tamped, and kept moist for a week. After the blocks are laid, cover all surfaces exposed to the fire with a solution of one-half pound of soap, one pound of slaked lime, applied with a flat brush. This solution will fill the pores and prevent discoloration.

Concrete for Hearths. The ordinary method of construction for work of this kind is to build a brick arch for the foundation, upon which the concrete is placed. This insures perfect fire protection. The top may be dressed with a trowel, or it may be finished with tile, as some prefer. The concrete for a hearth should consist of one part Portland cement, three parts sand, and five parts crushed stone or gravel, all thoroughly mixed.

ENGLISH SYSTEMS OF FLOOR CONSTRUCTION

In this connection, some valuable information will be found in the following account of methods of floor construction employed in England:

Within the last twenty years, a very great number and variety of fire-resisting floors have been adopted and patented in England. Some have died a natural death for want of fitness for the object in view; others have been kept alive by means of advertising and persistent push, irrespective of merit. In the course of time matters will adjust themselves, and Darwin's theory of the survival of the fittest probably applies to concrete construction as well as to the development of living species. It must not be assumed that the floors illustrated are suggestive of the best. They are intended only to show the progress and changes made in the methods of construction, the materials employed for the purpose, and typical examples of those more generally known and in use at present.

The opinion at one time was that if the materials of which floors and other parts of buildings were made were incombustible, they were fireproof; their behavior when simultaneously exposed to intense heat from the under side and the application of streams of water playing on the upper side, were not taken into account. It has nearly always been the case that the bottom

flanges of iron and steel beams have been quite
exposed, or at most covered with a thin coat of
plaster. As the iron joists must, in a large fire,
under these conditions, attain a high temperature in a short time, the heavy weight would
cause them to bend; and the lower half of the
concrete floor, already weakened by heat, would,
as a result, be subject to an increased tensile
strain. The heat would render the bottom
flanges weak at the time when the strain upon
them was greatest; and a general collapse was

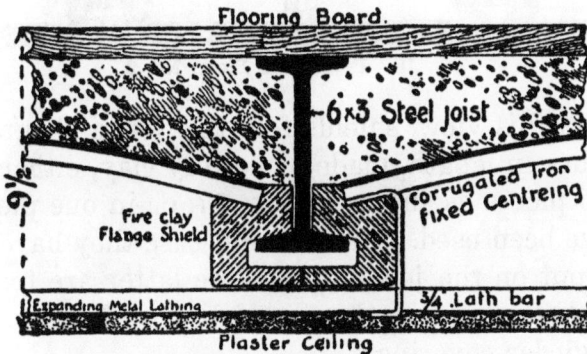

Fig. 17. Early Fireproof Construction—Steel Joists Protected by
Clay Shield.

often the result, causing fireproof floors in their
early days to be avoided in buildings on fire.
Another cause would be the expansion and sub-
sequent contraction of the concrete mass; for, if
the floor was in inseparable sections, each of
many superficial feet, it would probably part at
its weakest places through the heat causing it
first to expand and then contract as it cooled.

What it amounts to is that to render fire-resisting floors as much so as possible, every portion of steel or other metal construction should be encased in or covered with some material which both resists heat without breaking up, as far as practicable, and is also a bad conductor.

Fire clay or pottery clay will effect the purpose to a great extent if some way is adopted to afford means of expansion and contraction.

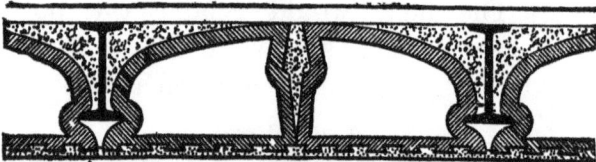

Fig. 18.   Ferguson's Floor, No. 2.

With this view, shields for steel joists about 12 inches in length, made of pottery clay, either in two pieces as shown in Fig. 17, or in one piece, have been used. In the latter case they have to be put on the joists before the latter are fixed, for both floors, arches, or slabs and for column or girder coverings.

Another writer says: "For certain purposes the porous terra-cotta lintel may be useful and economical; but, generally speaking, concrete is found to be more practical and constructionally more sound, and, all things considered, less costly. As for dense terra-cotta for floor construction, or the protection of metal work, its application is scientifically and practically wrong; in fact, terra-cotta is distinctly dangerous from the fire point of view. The complete disappearance of dense terra-cotta in floors and protective coverings intended to be fire-resisting, is only a matter of time."

The expressions "porous" terra-cotta and "dense" terra-cotta are intended to apply, the latter to material as used for building purposes in the ordinary way, while the "porous" terra-cotta consists of common fire clay, in which sawdust, shavings, charcoal, tree bark, or other combustible materials in a fragmentary state, are mixed, and which, being consumed at the time the blocks are being burnt in the kilns, renders them porous. The practice was known in this country [England] many years ago, and it is said was introduced to enable nails to be driven for the fixing of joinery or timber work.

It is commonly supposed to be an American invention, by reason of having been so largely adopted there for floors, etc. On the other hand, although concrete will stand a severe fire without collapsing, there can be no doubt that exposure to high temperature tends to weaken it considerably; and when this occurs, both concrete walls and floors should be tested before they are passed as safe. The erection of enormous smoke shafts with concrete is an American invention, and is an experiment worth watching.

Smoke flues of chimneys of ordinary dwellings, where constructed of river gravel concrete, usually give evidence, when swept, of some dislodgment of surface concrete; and these high chimneys are said to be made of concrete composed of Portland cement and Thames sand, the latter being the detritus arising from flint strata in the chalk formation.

A type of hollow fire clay tile construction floor is shown by Ferguson's, Fig. 18, which has

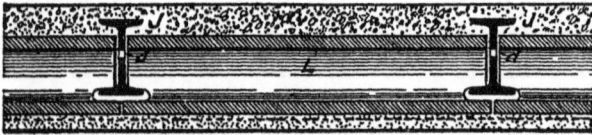

Fig. 19. Fawcett's Floor.

been in use about fifteen years. The steel joists are fixed two or three feet apart; the tubes are slotted to pass under the joists, and the spandrels filled in with concrete in the usual way. Unlike most other floors of this character, there are no skewbacks. The depth of the floor tubes

Fig. 20. Fawcett's Floor.

is $7\frac{1}{2}$ inches in the center, and they weigh about nineteen pounds per square foot. Dovetail grooves, as in Bunnett's of 1858, give a key for the plaster.

In Homan's system of fire resisting tile construction—or rather one of them, for the Homans have taken out enough patents during the last forty years, one would think, to cover the whole field of fire protection—the fire clay lintels are slotted at the ends. as in other systems, to pass under, and the soffits are also grooved for a key for the plaster. Steel joists are used, and tubular lintels, with concrete filling in on the lintels. The floor finish may be of cement, or of wood blocks or flooring nailed to the concrete.

Fawcett's floor (Figs. 19 and 20) is a class of floor similar to Homan's, the difference being only in the section of the lintels. JJ are steel joists; DD, holes in joists for ventilating purposes; LL, fire clay lintels; F, concrete filling which takes a bearing on the flanges of the joists, relieving the weight on the lintels. Fawcett's floor was patented in 1888.

The Kleine floor is formed with hollow burnt

Fig. 21. Frazzi Floor.

clay blocks laid in courses. Between the courses, light steel bars on edge are built in as the work

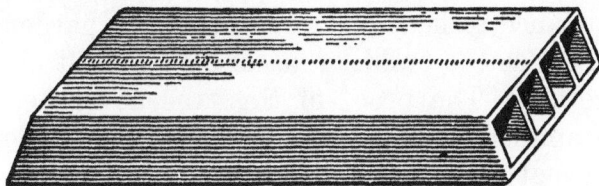

Fig. 22. Frazzi Floor.

proceeds, the ends of which rest on the walls or beam supports, concrete being filled on top to the depth required, and the blocks require a temporary platform beneath.

The Frazzi floor (Figs. 21 and 22) has fire clay lintels with skewbacks of the same materials, and concrete filled over the top in the usual way.

FIREPROOFING

In the modern improvements of fireproof building material, masonry plays an important part. Not infrequently it is considered of secondary importance by builders and structural engineers; but in the various large fires of our cities the condition of the masonry after collapse of a building has determined to some extent the relative value of the mortar employed. In the Baltimore fire, a good many collapsed buildings indicated that inferior mortar was used; and the weakness of the buildings was due to this, fully as much as to the quality of bricks, stone, or tiles.

It has become an axiom of modern building laws that mortar should approximate the same resistance to heat and pressure when hardened and set as the building materials which it binds together. The proof of this was never better exemplified than in the construction of modern fireproof buildings of the first class.

Problems for the Mason. In the employment of cement and other mortars for building fireproof structures of the highest grade, the mason has problems to solve that require the very best of ability. The functions of mortar, as have been described, are to unite building materials in such a way as to give the maximum surface of friction to the blocks or bricks, with the minimum amount of mortar. In other words, too much mortar is as bad as too little. Mortar, when employed between bricks or stones, sets much harder than when tested alone. This makes it easier for the expert mason to supply cement mortar which will set as hard as the building materials used. Unless mortar thus forms a compact wall as hard as the blocks or bricks, it becomes the weakest part of the structure. In fireproof buildings, many defects have been found in the mortar, simply through the neglect of this point. In the old buildings where little attempt was given to fireproofing, it was sufficient that mortar should be strong and adhesive.

But in modern buildings the mortar must be fireproof as well as hard as the building materials. It must resist temperature of a very high degree, and also refuse to disintegrate and crumble when a stream of water is played upon it. The fireproof qualities of the best Portland cement are well known, and in this we have the basic principle of a cement mortar which will resist fire. But much depends upon the quantity and quality of the cement, the methods of mixing, and the skill of the masons.

In regard to the first, we find that cements show a variation in grades according to the fineness of grinding and the method of burning. Even the coarse parts of the best Portland are inert and do not adhere to sand or broken stone when moistened. It is thus necessary that grinding should be perfect, as well as burning; and to secure the best results, very uniformly powdered Portland cement should be used. A too large mixture of sand with the cement to form concrete for binding purposes, weakens the fire-resisting quality of the mixture. Likewise mixing requires expert workmanship, for a too free use of water weakens the concrete to a low point of adhesiveness.

In building walls of fireproof buildings, masons accustomed to handling tiles and burnt bricks of a high grade are desirable. A good many fireproof buildings have had their weakness intensified by bad masonry. Unless all the spaces between the tiles or blocks are completely filled with fireproof cement mortar, the heat from a fire may penetrate through these air spaces to the ironwork inside or even to the interior of the building. If the weakness of the walls depends upon the masonry work, it is evident that the very best skilled labor is none too good for laying the courses of tiles, bricks, or concrete blocks.

The relative expansion and contraction of mortar with heat and cold is a point that interests the mason of to-day fully as much as any question in modern building operations. The cement mortars of the highest grade are relatively stronger in their adaptation to the changing effects of heat and cold in solid walls than any of the cheaper grades. The proper crystallizing and hardening of Portland cement depends upon the mixing and amount of water used; but in using such mortars with tile and porous terra-cotta, any excess of moisture is generally absorbed by the building materials. This leaves the mortar in the very best condition for setting permanently. It also accounts for the firmer setting of mortar when placed between two bricks or tiles than when left exposed on a surface. After the proper crystallizing of the cement, the quicker it sets the better must the results prove.

Destruction by Heat. The destruction of mortar by heat is due more to the action of the fire upon the sand and broken stone incorporated with it than upon the cement itself. This has been demonstrated many times in tests and in actual experience. Where there is an excess of sand or broken stone mixed with the cement, there is sure to be a lower degree of fireproofing in the mortar. This fact is not always sufficiently emphasized. In the effort to cheapen the cost of the mortar, workers are tempted to add sand too freely in proportion to the amount of cement used, but not to such an extent as to actually weaken the strength or adhesiveness of the mortar. Those not accustomed to dealing with fireproof work may not realize the enormity of their sin. The fact that they must consider the fireproof feature of their mortar as well as its strength, should indicate to them the duty of maintaining the exact proportions of sand and cement. **Concrete properly made is absolutely fireproof, provided it is thoroughly dried and contains no moisture.**

Hardness and Resistance. It has become an axiom of modern building laws that mortar should approximate the same resistance to heat and pressure when hardened and set as the building materials which it binds together. This in effect produces one compact mass in a wall. Mortar, when employed between bricks, stones, or building blocks, sets much harder than when tested alone. This fact makes it easier for the worker to supply cement mortar which will set as hard as the building materials used.

SYSTEMS OF HOUSE CONSTRUCTION

Among the systems of house construction recently evolved may be mentioned the following:

The Edison House. Thomas A. Edison, the inventor, is perfecting a system of construction which is popularly known as the **poured cement house.** By this system he intends to use cast-iron moulds which will be so arranged that the entire house, a two-story structure designed for working people, can be poured in six hours after the moulds are set up. See Plate 4.

This house is for one family, with a floor plan 25 by 30 feet. It is intended to be built on lots 40 by 60 feet, giving lawn and small garden room.

FERRO-LITHIC PLATE AS USED IN ROOFING.

Concrete is applied above, and plaster underneath, as shown in central figure.

PLATE 3—CEMENT CONSTRUCTION.

The front porch extends 8 feet, and the back porch 3 feet.

On the first floor is a large front room 14 by 23, by 9½ feet high, intended as a living room; and a kitchen in the back 14 by 20, by 9½ feet high. In the corner of the front room is a wide staircase leading to the second floor.

The second floor contains two large bedrooms, a wide hall, and a roomy bathroom (7 feet 6 inches by 7 feet 6 inches, by 8 feet 2 inches high). The third floor has two large rooms.

Each room has large windows, so that there is an abundance of light and fresh air.

The cellar, 7 feet 6 inches high, extends under the whole house, and will contain the boiler, wash-tubs, and coal bunker. The main room, as well as the outside of the house, will be richly decorated.

The decorations are cast with the house and therefore form part of the structure, instead of being stuck on, as is done at the present time.

The Graham System. Owing to the growth of our cities and the massing together of people, the demand for fireproof buildings has been steadily increasing. Engineers and builders realize that every owner wants his building to be fireproof; but up to the advent of concrete, all methods of fireproof construction have been expensive and almost prohibitive, except for the highest type of building where excessive rentals can be obtained.

The more common methods of reinforced concrete construction, although fireproof, and less expensive than steel and tile fireproofing, are not extensively applied to residence or apartment houses, nor other buildings of moderate size.

The difficulties attending other forms of fireproof construction seem to have been largely overcome in the form of building invented by Mr. George M. Graham, which is an entirely new combination of steel tubing, wire, malleable fittings, and concrete. With the exception of piers, the concrete is not depended on to carry any of the load, but is used only as a stiffener or body to the building. The entire framework of steel tubing can be erected before the concrete work is started, making it possible to inspect the position and quality of steel, and to erect a building in a much shorter time—it is claimed—than is required by any other form of construction. Also, no forms or centering are required, which gives this method of concrete construction an added advantage. The walls and floors are hollow, which reduces the weight of the building to the minimum and affords perfect insulation. The strain on the floors is carried by wire in tension, which is an economical way of using steel, and the tests have shown that an equal amount of steel used in this way makes a floor of nearly double the strength of any other form of construction. The walls, floors and partitions form one integral mass, so that the building is absolutely vermin proof and indestructible, even by earthquake. As every partition, floor, and ceiling is interwoven with wire, it is practically impossible for cracks to develop. The building is absolutely fireproof, and costs very little more than the present form of brick walls, wood floors, and partitions, which are so highly inflammable. All steel and wire are encased in cement, which prevents corrosion or rust. The exterior is of cement mortar, which permits of any finish or form of ornamentation desired.

Hollow Wall Method. A highly developed type of monolithic reinforced hollow concrete wall is shown in the work done by an inventor at Petoskey, Mich.

In this method of building, the stone is moulded upon the wall. The moulds travel on a track which is attached to an elevated scaffold. They are 12 inches in height; and as rapidly as each course is completed, the scaffold is simply raised 12 inches, which operation also raises the track and moulds, and a new course is started. This same movement is repeated until a story is erected, when the joists are placed, and the entire apparatus is raised to that floor and a new story added; consequently the same apparatus is sufficient to build any number of stories in height.

Between each 12-inch course and its adjacent courses, the walls are tied together with steel ties which reinforce and strengthen the walls, lengthwise and crosswise, thus making the en-

tire walls one complete network of steel reinforcement.

The walls being monolithic throughout the
building, they are not affected by fire, acids,
gases, water, climatic changes, etc. Window and
door sills are moulded any size or length, wherever they come in the structure, and at once
become a part of the one solid stone which the
building is when completed.

The entire work being done upon the walls,
special moulds can be used on the outside, making floral or any other artistic designs one
may desire. While the stone is green, it can be
marked off representing block work or stone in
any length. The natural concrete is always
more beautiful and lasting than any imitations
of other work and material.

The Aiken Method. A novel method of
building construction has recently been carried
out in connection with structures erected for the
State militia at Camp Perry, Ohio, and at other
places. The use of concrete in the composition
of the walls has permitted them to be practically
completed before being placed in position. See
Plate 5.

The process of constructing a wall was as follows: First a platform of 2-inch lumber was laid
across steel beams about 4 feet apart, these
beams being supported by jacks. The platform
was about 3 feet from the ground, and lay inside
the limits of the proposed building. Four-inch
boards were set up on the four sides to complete

the form. On the platform were placed the window-frames and the reinforced concrete cornice, which was moulded in 6-foot sections, 3 feet wide. In this case, special ornamental window-caps were required, and these were cast separately and placed in their proper positions on the platform. Then concrete made of one part cement, one and one-half to two parts sand, and four parts crushed stone, was poured upon the platform. After about 2 inches of concrete had been laid, twisted steel rods for reinforcements were placed in both directions, 6 inches apart, and the balance of the concrete was poured on. The wall was made 4 inches thick. As a facing, a cement mixture of one part white cement to one and one-half parts white sand was laid on the surface.

The work was allowed to stand forty-eight hours to give the material time to solidify, when preparations for lifting the wall to its permanent position were made. This was a comparatively simple task, most of the power being furnished by a 5-horsepower engine. It was connected by belting with the shaft under the platform operating the jack-screws, and slowly the wall was tilted into position. The platform supports were so placed that the foot of the wall swung to its position on the foundation at precisely the right line; and when the wall had assumed a vertical position, every line was plumb. Five or six wood props braced to the window-frames held the wall in position, and the platform was taken away from the back, and swung about for the construction of the next wall, at right angles to the first. This operation was repeated until all the walls were up. The reinforcing rods were set to protrude at the edges of the walls; and when all the walls were in position the rods interlocked at the corners of the structure. They were twisted together, and an 8-inch board, the only falsework used in the construction, was placed inside the corner. Here concrete was poured in, a joint made on the outside corner, and the two walls thus bound together.

Apparently this system could be utilized in constructing walls of large dimensions, provided the adjustable framework for supporting the wall is of sufficient strength to give equal resistance to all portions of the load while being raised.

BLOCKS WITHOUT FACING

The appearance presented by the concrete building block is a subject that has engaged the serious attention of men interested in the growth of the industry for a number of years. And it is to be said that some of the problems growing out of the use of these blocks are being solved— and by friends of the block. To the average man, the block is satisfactory in appearance in some one or more of the forms or designs now in vogue. It is safe to say that to the majority of builders satisfaction can be secured by some combination.

A writer of wide practical experience in the manufacture and use of concrete blocks, gives us the benefit of his experience and observations in the following paragraphs:

The critical home-builder sometimes makes objection to the sameness of the concrete block. This is really the only objection that is urged against the block. It is to the man who objects to the sameness that this article is addressed. We are confident we have a plan of construction that solves the problem for him and still enables him to employ concrete blocks in building. The writer heard an intending home-builder say a few days ago: "I would build my house of concrete, but it costs too much for forms and I can't afford it; nor will I use blocks."

The old adage "if thy right hand offend thee, cut it off" applies here. If the face of the block offends you, cut it off. In any case where it is desired to secure solid effects in a wall, the blocks may be made **without a face** and cast with a **rough surface**. Then the wall may be laid with them, and a finish of any desired style may be given by a plaster coat of cement mortar.

If you stop to consider a wall of this description a minute, you will readily see the economy of it. A wall may be laid up with blocks in this manner and plastered in any desired finish, without the use of a single stick for forms. The joints of the blocks, too, provide for the expansion and contraction that would have to be figured in a monolithic wall. The entire effect will be that of a monolithic wall; but in reality it will be made of hollow blocks and answer every demand of monolithic construction.

Now this suggestion is made only for the man who objects to the sameness effects of concrete blocks with faces. It does not involve at all the elimination of the face block from the building market, for the artistic face block is here to stay, and thousands of builders will continue to find pleasing combinations and designs for buildings. To the builer, the contractor, or the architect who has presented to him an objection, the plan here outlined offers an argument and solution.

As to the method of making blocks like those described, it may be said that only the rough material usually employed need be used. The surface that is to be plastered should be rough in order to provide a holding surface; and when the blocks are laid in the wall, the joints should be left free of mortar for half an inch. Then, before the plaster coat is applied, the wall should be well sprinkled so that the mortar will not be robbed of the water it needs during the process of setting.

And now as to the use of cement mortar on brick, it is not an uncommon thing to see a concrete building faced with brick of special color and terra-cotta for different effects. On the other hand, it looks sometimes as if the reverse of this practice were coming into vogue, and, if not now, it soon will be no uncommon thing to see brick buildings faced with concrete or some other product that belongs to the concrete class.

The picture shown in Plate 5 **A** is a snapshot of the new Whitesides Bakery in Louisville, which is built of brick and is faced with cement mortar. The snapshot shows the men at work on the side of the building applying this coat of mortar, which is of a particular kind. The mixture is made of cement and crushed tile. The crushed tile includes a variety of colors: blue, gray, white, terra-cotta, etc. As it was piled on the ground ready for use, it looked, from a distance, like small gravel taken from a stream, but close investigation showed its real nature. It is applied to the face of the brick wall something like plaster, and then a spray hose is turned on it to wash off whatever coating of smooth cement may form on the outside, so that the tile particles may show as a rough wall. The foundations and corners shown in the picture are made of stone, and there are stone trimmings through the building which show up nicely along with the gray finished wall. The roof on the tower, after being completed, was covered with red clay tiling, and there was some terra-cotta ornamental effect in trimming, so that altogether the structure presents a very artistic appearance.

The idea suggested by a study of this building to the writer was that many of the older brick buildings might be plastered on the outside with concrete in this manner, and suitable trimmings added so that they would present a much better appearance than can be obtained by painting the brick, or by putting a smooth plaster of cement and sand over it.

COLOR OF CONCRETE BLOCKS

Outside of all considerations of form and effect, the matter of color is not the least important of the properties of the concrete block. It is upon this more than anything else that the marketing of the block depends. It is, therefore, important to secure uniformity of color, and an effect that will be pleasing to the eye. This requires constant study to attain in its perfection, and is one of the greatest problems of the business. For colored blocks, dark-colored cements may be used to the best advantage, but their color should be uniform. For lighter blocks, cement of lighter color must be used. Cement gets lighter as it is ground finer, and only finely ground cement should be employed for this work. Cements entirely free of manganese and iron would be white; and the color grows dark as the percentage of iron and manganese increases. Most cements, however, have more iron than manganese, and the color, therefore, is due to the iron.

Sulphate of lime, which is always added to cement to regulate the set, is responsible for the white efflorescence of blocks. The salt is sol-

TABLE II

Coloring of Cement Mortar

DRY MATERIAL USED.	WEIGHT OF DRY COLORING MATTER TO 100 LBS. CEMENT				COST OF COLORING MATTER. PER LB. CENTS.
	½ lb.	1 lb.	2 lbs.	4 lbs.	
Lamp Black	Light Slate	Light Gray	Blue Gray	Dark Blue Slate	15
Prussian Blue	Light Green Slate	Light Blue Slate	Blue Slate	Bright Blue Slate	50
Ultramarine Blue		Light Blue Slate	Blue Slate	Bright Blue Slate	20
Yellow Ocher	Light Green	Pinkish Slate		Light Buff	3
Burnt Umber	Light Pinkish Slate		Dull Lavender Pink	Chocolate	10
Venetian Red	Slate. Pink Tinge	BrightPinkish Slate	Light Dull Pink	Dull Pink	2½
Chattanooga Iron Ore	Light Pinkish Slate	Dull Pink	Light Terra Cotta	Light Brick Red	2
Red Iron Ore	Pinkish Slate	Dull Pink	Terra Cotta	Light Brick Red	2½

uble, and is carried to the surface of the blocks during the curing process.

In coloring artificial stone to a **gray,** the use of one pound of Germantown lampblack mixed with cement dry, and one pound of salt previously dissolved to every ten gallons of water, greatly assists to waterproof the product, but does not make an absolutely waterproof stone, although increasing the lampblack makes the stone less absorbent and affects the durability.

Black stone is produced by adding peroxide of manganese to the cement in the proportion of twelve to fifty pounds per barrel of cement. The amount is governed by the color of the sand and cement, or by adding from two to four pounds of excelsior carbon black to each barrel of cement. The manganese in a measure prevents the absorption of water, but the excelsior carbon does not; the first reduces the strength, and the other has little or no effect upon the durability or strength of the product.

Blue. The use of any waterproofing in making black stone will discolor the surface. In producing waterproof blue stone, the best results are obtained by using five pounds of ultramarine blue, one pound of pulverized alum, and one pound of soda mixed dry with the cement. This produces a very sound product, less subject to moisture than perhaps any natural product.

Red artificial stone, made either of oxide of iron or Pompeian red, will not mix well with waterproofing compounds, and it is the opinion that reduction of the absorbing qualities can be done only after the colored product has hardened.

Brown or Buff. In making brown or buff stone, ocher, which in a measure prevents moisture, is employed; but ocher is detrimental to strength.

Table II, based on experiments made by L. C. Sabin, an authority on the subject, shows the color results obtained from a dry mortar, the mortar containing two parts sand and one of cement.

Mix Coloring with Cement. In the production of colored blocks, the coloring matter should be mixed with the cement so that in effect it will be colored cement that will be mixed with the sand. This method assures a thorough coloring of the block, and a uniform shade throughout.

CONCRETE BLOCK INDUSTRY

No other department of the cement industry has so felt the need of standard specifications and uniform instructions as has the manufacture of cement blocks.

There is to-day a large and growing demand for this material, and its general and almost unlimited use is retarded only by lack of confidence on the part of architects, builders, and residence owners who see only the wretched results that attend the efforts of the misinformed and inexperienced, and overlook the splendid possibilities of this form of construction in the hands of skilled and experienced operators.

In considering the requirements that cement blocks should meet as a structural material, we must take into account the use to which they are to be put.

We have in brick classification, the terracotta brick, mud brick, and dry-pressed face brick, and the hard-burned, medium, and light common brick—all of which find extensive and legitimate use, and yet vary widely in strength, fireproof qualities, and appearance.

The granites, limestones, sandstones, and marbles are generally accepted in first-class construction, and yet differ greatly in weather- and fire-resisting qualities.

Lumber, of course, is very combustible; and

yet the different varieties show marked contrast in strength, durability, and fire-resisting qualities, and we have still to learn of any municipal requirements stipulating the kind of lumber for building construction.

With these facts in mind, is it not fair to ask that some latitude be granted in the manufacture and use of cement blocks?

If an owner in most localities chooses to build the outside walls of his factory or residence of light-burned common brick, showing an absorption of 30 per cent water, who is there to raise objection? In fact, the average so-called hard-burned brick will absorb 20 to 22 per cent water, and will pass muster under most municipal and architects' requirements; yet our leading municipal specifications require that cement blocks shall not exceed 15 per cent absorption, regardless of the use to which they are to be put.

Uses of Cement Blocks. Cement blocks may be properly used in substitution of other materials for:

1. Foundations.
2. Exterior and superstructure walls carrying weight.
3. Curtain walls, exterior and interior.
4. Fire walls and partitions.
5. Veneering.
6. Retaining walls.
7. Cornice, trim, and ornamental work.
8. Filler blocks for floor slabs.
9. Chimney flues, etc., etc.

In this variety of work it is at once seen that

uniform quality—and the highest quality—is now required.

Experience in the use of other materials has taught us to recognize, practically without repeated or preliminary tests, the quality of most materials for which cement blocks are substituted; and this fact alone gives these older materials an advantage over the newer.

Commercial, local, and natural causes, however, are calling for the more extensive use of cement blocks. This demand will increase as our manufacturers of cement blocks gain experience, and through the observance of rational building requirements. It is of prime importance to every city and town in this country, having a building code, that it should recognize and include cement blocks as a building material.

The writer of the specifications herewith submitted, Mr. E. S. Larned, C. E., as Chairman of the Committee on Tests of Cement and Cement Products of the National Association of Cement Users, recommended in a report, that a Specification Committee be appointed by the Association to draw up a standard specification and uniform instructions covering the manufacture of cement blocks, with the hope that this form, when prepared, might be offered to all the cities and leading towns in the United States for adoption.

As a basis upon which to consider the matter of standard specifications and uniform instructions, his suggestions included the following in part:

Cement. Only a true high-grade Portland cement meeting the requirements and tests of the standard specifications of the American Society for Testing Materials shall be used in the manufacture of cement blocks for building construction.

Unit of Measurement. The barrel of Portland cement shall weigh 380 pounds net, either in barrels or subdivisions thereof made up of cloth or paper bags; and a cubic foot of cement packed as received from the manufacturer shall be called 100 pounds of the equivalent of 3.8 cubic feet per barrel. Cement shall be gauged or measured either in the original package as received from the manufacturer, or may be weighed and so proportioned; but under no circumstances shall it be measured loose in bulk, for the reason that when so measured it increases in volume from 20 to 33 per cent, resulting in a deficiency of cement.

Proportions. Owing to the different values of natural sand or fine crusher screenings for use in mortar mixtures, due not only to its mean effective size, but also to its physical characteristics, it is difficult to do more in a general specification than fix the maximum proportions of good sand that may be added to cement.

Sand, or the fine aggregate, shall be suitable silicious material passing the one-fourth-inch mesh sieve, and containing not over ten per cent

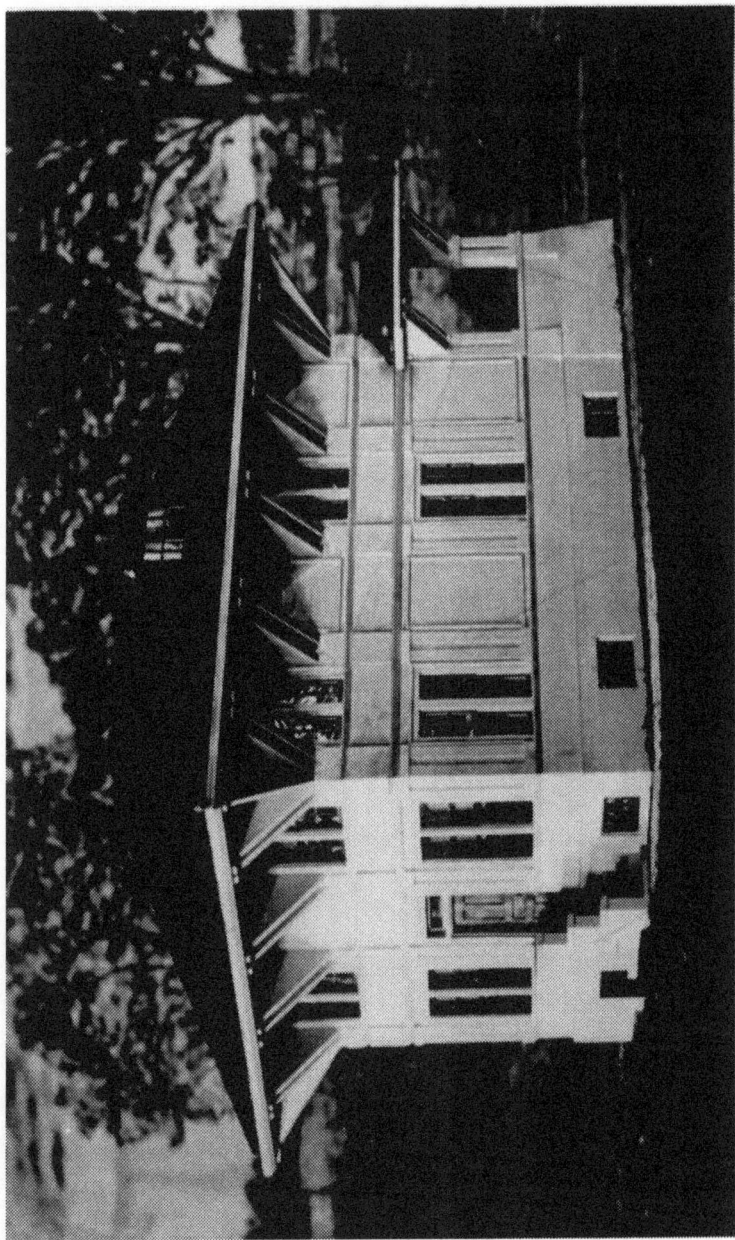

PLATE 4—CEMENT CONSTRUCTION.

THE EDISON POURED HOUSE.

UNIT-SECTION OF VESTIBULE WALL AT CAMP PERRY, OHIO.

PLATE 5—CEMENT CONSTRUCTION.

of clean, unobjectionable material passing the No. 100 sieve. A marked difference will be found in the value of different sands for use in cement mortar. This is influenced by the form, size, and relative roughness of the surface of the sand grains, and the impurities, if any, contained.

Only clean, sharp, and gritty sand, graduated in size from fine to coarse and free from impurities, can be depended upon for the best results. Soil, earth, clay, and fine, "dead" sand are injurious to sand, and at times extremely dangerous, particularly in dry and semi-wet mortars; and they also materially retard the hardening of the cement. An unknown or doubtful sand should be carefully tested before use, to determine its value as a mortar ingredient. Screenings from crushed trap rock, granite, hard limestone, and gravel stones are generally better than bank sand, river sand, or beach sand in Portland cement mortars (but not so when used with natural cement, unless the very fine material be excluded).

So-called clean but very fine sand has caused much trouble in cement work, and should always be avoided, or, if impossible to obtain better, the proportion of cement should be increased. Stone screenings and sharp, coarse sand may be mixed with good results; and this mixture offers some advantages, particularly in making sand-cement blocks.

For foundations or superstructure walls exposed to weather, carrying not over eight tons per square foot, the maximum proportion shall not exceed four parts sand to one part cement. This proportion, however, requires extreme care in mixing for uniform strength and will not produce water-tight blocks. We recommend for general work not over three parts sand, if well graded, to one part cement, and the further addition of from two to four parts of clean gravel stones passing the three-fourths-inch sieve and retained on a one-fourth-inch mesh sieve, or clean screened broken stone of the same sizes. These proportions, with proper materials and due care in making and curing, will produce blocks capable of offering a resistance to crushing of from 1,500 to 2,500 pounds per square inch at twenty-eight days.

(For the best fireproof qualities limestone screenings or broken sizes should be excluded, but otherwise are all right for use.)

Where greater strength is desired, particularly at short periods, from two to six weeks, we recommend the proportions of one and one-half to three parts gravel or broken stone of sizes above given. Blocks made of cement, sand, and stone are stronger, denser, and consequently more waterproof than if made of cement and sand only, and are more economical in the quantity of cement used.

Mixing. The importance of an intimate and thorough mix cannot be overestimated. The sand and cement should first be perfectly mixed dry and the water added carefully and slowly in proper proportions, and thoroughly worked into and throughout the resultant mortar. The moistened gravel or broken stone may then be added, either by spreading same uniformly over the mortar, or by spreading the mortar uniformly over the stones; and then the whole mass shall be vigorously mixed together until the coarse aggregate is thoroughly incorporated with and distributed throughout the mortar.

We recommend mechanical mixing wherever possible, but believe in the thorough mixing of cement and sand dry, before the addition of water; this insures a better distribution of the cement throughout the sand, particularly for mortar used in machine-made blocks of a semi-wet consistency. For fine materials, such as used in cement blocks, it is necessary that the mechanical mixer be provided with knives, blades, or other contrivances to thoroughly break up the mass, vigorously mix the same, and prevent balling or caking.

Curing. This is a most important step in the process of manufacture, second only to the proportioning, mixing, and moulding, and, if not properly done, will result either in great injury to or the complete ruin of the blocks. Blocks shall be kept moist by thorough and frequent sprinkling, or other suitable methods, under cover, protected from dry heat or wind currents for at least seven days. After removal from the curing shed, they shall be handled with extreme care, and at intervals of one or two days shall be thoroughly wet by hose sprinkling or other convenient methods. We recommend curing in an atmosphere thoroughly impregnated with steam. This method serves to supply needed moisture, prevents evaporation, and in some measure accelerates the hardening of the blocks.

We view with distrust, in the present knowledge of the chemistry of cement, any artificial, patented, or mysterious methods of effecting the quick hardening of cement blocks or other cement products. If such method be proposed, it should be thoroughly investigated by competent authority before use.

Time of Curing. This is also most important in its effect upon the industry, and is directly and vitally influenced by the following conditions:

1. Quality, quantity, and setting properties of the cement used.

2. Quality, size, and quantity of the sand or fine aggregates used.

3. Amount and temperature of water used.

4. Degree of thoroughness with which the mixture is made.

5. Method of curing, weather conditions, and temperature.

6. Density of the block as affected by the method and thoroughness of tamping, or by the pressure applied.

Before fixing the minimum permissible time required in curing and aging blocks, it is well to consider the important effect of additions of sand upon the tensile strength of cement mortar.

The following tabulation has been interpolated from the diagram of cement mortar tests prepared by Mr. W. Purves Taylor, of the Philadelphia Municipal Laboratory.

The results of the neat tests and the 1 to 3 mortar tests (that is, one part cement to 3 parts crushed quartz by weight) are averaged from over 100,000 tests, while the other results are based on from 300 to 500 tests.

Tensile Strength in Pounds per Square Inch of Portland Cement Mortar

PROPORTIONS	7 days	28 days	2 mos.	3 mos.	4 mos.	6 mos	12 mos.
Neat cement........	710	768	760	740	732	758	768
1 to 1 mortar.......	590	692	690	680	680	685	695
1 to 2 mortar.......	370	458	460	455	453	458	460
1 to 3 mortar.......	208	300	310	310	310	310	308
1 to 4 mortar.......	130	210	230	230	230	232	232
1 to 5 mortar.......	80	150	185	195	195	195	197

It must also be kept in mind that these results are obtained under practically uniform and theoretically correct conditions, in the amount of water used, thoroughness of mixing, and moulding and storage of samples until tested.

Comparing the results at 28 days, it is apparent that the 1 to 5 mortar has only 71 per cent of the strength of the 1 to 4 mortar, and but 50 per cent of the strength of a 1 to 3 mortar. The 1 to 4 mortar has but 70 per cent of the strength of a 1 to 3 mortar, and 46 per cent of the strength of a 1 to 2 mortar.

The ratio of compressive strength to tensile strength is not quite constant for all periods of time, and for the several mixtures above given; but the compressive strength, or resistance to crushing per square inch, may be approximately obtained by multiplying the tensile strength given in the above table by the constant six (6). (Note—This would increase with the age of the mortar, and would be greater for good gravel or stone concrete than for the clear mortar of which a given concrete is made.)

In fixing the minimum time required for curing and aging blocks before use, due regard should be given to the proportions used. It is manifestly wrong in principle to require as long a period for a 1 to 2 or a 1 to 3 block as might seem necessary for a 1 to 4 or a 1 to 5 block; and it is obviously unsafe to attempt to use a block of lean proportions in as short a time as a rich mixture would gain the necessary strength.

This might be supposed to be met by fixing the minimum resistance to crushing of blocks (of all compositions); but it must be kept in mind that a very small percentage of the blocks used are tested, by reason of the expense, inconvenience, or lack of facilities.

The required minimum resistance to crushing of first-class blocks used for exterior and bearing walls should not be imposed upon blocks for minor and less important uses.

Marking. All cement blocks should be stamped (in process of making), showing name of manufacturer, date (day, month, and year) made, and composition or proportions used. The

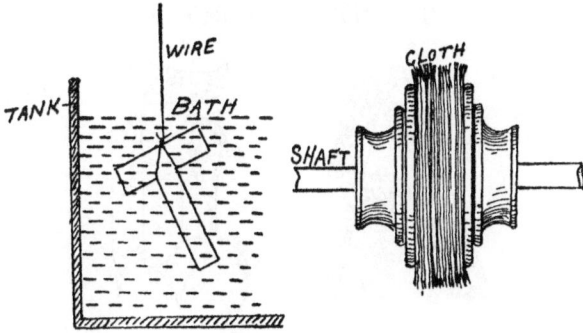

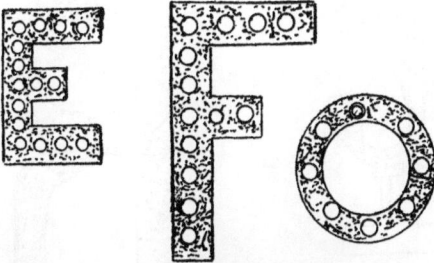

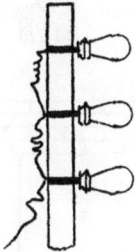

Fig. 23.   Making Cement and Other Letters for Advertising Pur
poses, Showing Electroplating Devices, etc.

103

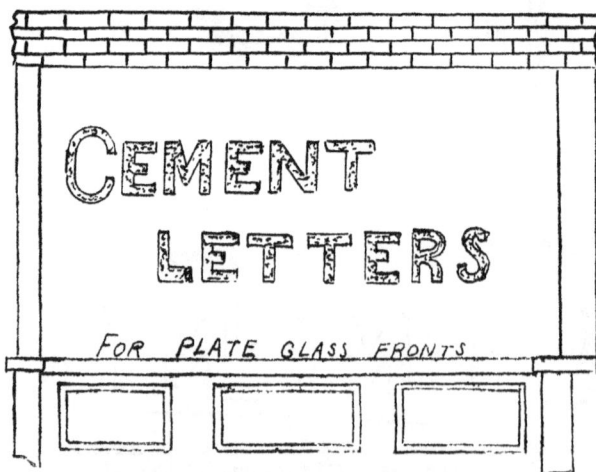

Fig. 24. Types of Cement Letters.

place of manufacture, methods, and materials should also be open to inspection by representatives of the Building Department, the architect, engineer, or individual buyer.

LETTERING WITH CEMENT

One of the branches of the cement business attracting attention at the present time is the manufacture of **letters** with cement. Cement letters have been used for years, but it is only recently that very much has been done in a practical way in this direction. The manufacture of cement letters has been restricted to the making of a few plain letters for special purposes. Sometimes letters of cement have been wanted to use on stone surfaces. Then, again, they are used on plate-glass fronts of stores, when properly finished. There are processes of coating the cement letters with finishing compositions by which a gloss is obtained. There are plating and nickeling operations for making the proper surface on cement letters. But as a general rule, the cement letter is preferred in the rough state. The genuine surface of the cement presents a novelty in itself, and this surface is attractive to many persons. The makers of cement letters are often called upon to mould letters for exhibition purposes in advertising. In fact, the cement letter advertising tablets are among the most thriving divisions of the cement letter industry. The accompanying sketches will give one an idea of the scope and service of the cement letter. The writer visited some of the cement works where letters were being made. The concrete block makers have in some cases put in cement letter-making devices as a side line of work.

While the plain letters are used to a larger degree than the fancy designs, you will observe some very unique effects in service. Letters constructed on the plan of the letter A shown in Fig. 24 are common. The letters of the old English type, as shown in the letter D, are popular. Then there are letters molded in form as in the G shown, and set upon a pin. The object of this combination is to align a number of letters on a certain level in the spelling out of a word or sentence. Then there are letters of the style shown in the R. There is an endless variety of patterns of letters, some with smooth, glossy surfaces, and others in mottled effect.

A CONCRETE SAFE

If concrete safes come into general use the owner will find it cheaper to build a new one than to move the old safe; and safes can be had in all sorts of inaccessible places by carrying a few small bags of sand and cement, and building the safe from the materials. The concrete safe is not burglar-proof; neither is the average steel safe; but a safe built of concrete sufficiently strong for all ordinary requirements against theft and fire.

A concrete safe was recently made in Seattle, by a concrete building constructor. One who saw it says the safe is 2 feet 4 inches square and 3 feet high. Walls and doors are 4 inches thick, reinforced with ¼-inch twisted steel, with the lock and hinges cast in the center of wall and door. The handle and castors were also cast in place. The construction is not specially difficult—anyone handy with tools can make one; and the iron parts can all be purchased, and are comparatively inexpensive. A few dollars and a little interesting work will provide a good, durable, substantial safe.

Concrete on the Farm

Throughout the greater portion of the country the pioneer days of hardship and forced economy are over. These conditions led to flimsy construction; but the time has come now when the farmer can use better materials in his constructive work. All that is needed is that he shall understand the value of concrete for this purpose, its adaptability, endurance, and permanency. When this material is generally adopted in this manner in the rural communities, this country will be on basis structurally with the old countries of Europe.

The many uses to which concrete may be put by the farmer to insure to him durability and that imperishable quality so much to be desired, are being recognized more and more. Certain well-defined rules in the construction of concrete work, based on the experience of others, make the handling of cement by even the inexperienced an easy matter, provided, of course, that the rules are adhered to.

Any material that assures to the agriculturist the certainty that what he builds will last, not only during his lifetime, but during that of his children and his children's children, is to his advantage. And when it is considered that he can employ cement in making his improvements at a cost not exceeding that which he would have to pay if perishable lumber or timber were used, it is evident that concrete must be adopted generally by the wise home-maker and the agriculturist.

Windmill Foundation. The most frequent complaint against wooden foundations for windmills is concerning the habit they have of rotting. Concrete construction will overcome this, and give to the windmill a solid foundation that will insure it against being blown over in storms.

Excavate four holes at the desired distance apart, $2\frac{1}{2}$ feet square and 5 feet deep. Build forms for the sides, and grease them properly. Fill the forms 2 feet deep with concrete, one part cement, three parts sand, and six parts of gravel or broken stone, of a jelly-like consistency, tamping well every four inches.

The holding-down bolts can be suspended from a frame over the top, care being taken to place them so that they will be in true position when the concrete is placed around them. They should be two feet long, with plates to resist the pulling strain. Fill the form with concrete flush with the top, and allow it to remain for several days before using. This will make a substantial anchorage for a steel tower. If a wooden tower is to be used, run projecting bolts up through the timber sills, and use large cast-iron washers under the bolts. The anchorage in this case should project at least six inches above the ground.

Concrete Sinks. Sinks made of concrete are considered as durable as iron. They may be made in any desired size, and reinforced with wire netting. They are cast in wood or plaster moulds. The mixture should be one part Portland cement and two parts fine crushed granite. A rebated hole should be provided for a trap. The thickness should be two to three inches.

CONCRETE TANKS AND CISTERNS

Concrete Water-Tanks. It is unquestioned that a tank built of concrete is not only more durable than one of wood, but is more sanitary. Many farmers have them installed in their barns, and tanks of this character answer the requirements in numerous manufacturing plants.

Two forms are required in the construction of a tank. One is needed for the moulding of the exterior, and a smaller one for the formation of the inner surface of the walls. Each form should be made of dressed boards without knots that will disfigure the surface of the concrete. For a tank of medium size, say 2 feet wide, 6 feet long, and 2 feet deep, the walls should be three inches thick. Allowance must be made therefore in the making of the two forms for a space of this width.

A flat surface perfectly smooth should be provided on which to place the forms. The smaller one is put in position first, as it is the one that will form the interior walls. It would be well, before placing the larger form, to measure carefully for the placing of it, by laying it on the platform over the smaller one and marking the

WHITESIDES BAKERY, LOUISVILLE, KY.

Brick Faced With Cement Mortar.

PLATE 5A—CEMENT CONSTRUCTION.

CONCRETE SIDEWALK, HORSE-BLOCK, AND HITCHING POST.

A CONCRETE WATERING TANK.

PLATE 6—CEMENT CONSTRUCTION.

corners, fastening the form so that it will not move when the concrete is placed. The inner form should be fastened also for the same reason. Both forms should be greased.

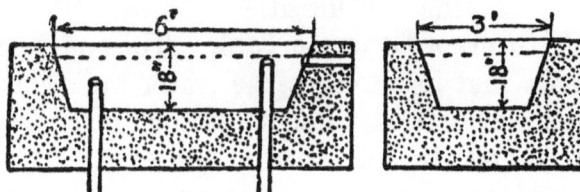

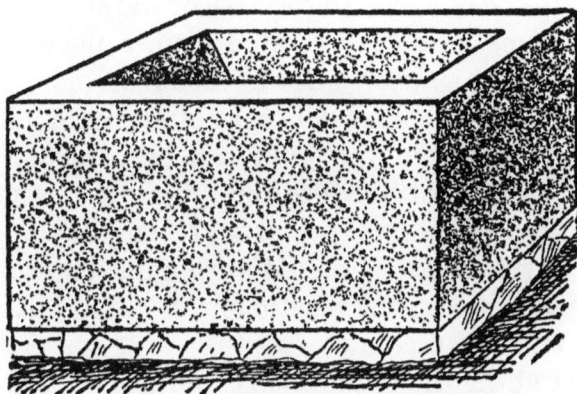

Fig. 25. A Small Concrete Tank.

The concrete should be prepared near at hand so that it can be placed quickly. The mixture should consist of one part Portland cement, two parts sand, and three of gravel or crushed stone that will pass a quarter-inch sieve. Be careful not to make the mixture too wet. Tamping in layers of three inches is recommended for this work, each layer being followed up quickly with the succeeding one so that the fixing may be uniform. The top is last leveled off and finished. Do not disturb the forms for a week or ten days. Then they may be removed, and the drying continued.

Waterproof Concrete Tank. The suggestion is made that in all cases the water tank should if possible be slightly flaring, so that, if the water in it freezes, the pressure of the expanding ice will be less than if the sides or ends were vertical. The point of this suggestion will be evident to the reader. In the case of water troughs, this shape is more desirable from a utilitarian point of view, as it affords easier access. The first factor to consider is that of strength. The waterproofing of the tank can be attended to later. The best mixture for a tank is one part Portland cement, three parts of sharp sand, and five parts of gravel mixed with sufficient water to make the cement plastic or "sticky," but not thin enough to pour.

First dig the foundations, and place the footings for the wall and floor. Then mix the concrete as follows: For the foundation up to within four inches of the floor line, use one part Portland cement, three parts sharp sand, and two parts gravel that will pass through a one-inch sieve, and four parts gravel that will pass a two-inch sieve. Mix thoroughly, and ram into position. If the tank is to be as large as 30 by 40 feet and 6 feet high, this course should be covered with a three and a-half inch course of one part Portland cement, three parts sand, and three parts gravel that will pass through a sieve of ¾-inch mesh. After this has hardened sufficiently, build the falsework with a plank on both sides, and put in all pipe connections before beginning to place the concrete for the walls, as any interference with the foundation after the walls are erected is not an easy matter to remedy. Mix one part Portland cement, three parts sharp sand, and five parts clean gravel that will pass a one-inch sieve, and just enough water to make it sticky, and ram hard into position. Leave all the falsework in position for at least three days, to allow the concrete to harden. After removing the plank, plaster the exterior surface with a mixture of one part Portland cement and two parts sharp sand that has passed through a screen of one-fourth-inch mesh. This will give a finished effect, and will hold if the surface is wetted before application. A waterproofing compound should be used in all the mixtures.

Inlet and outlet holes may be made by putting pieces of pipe in place before filling in the concrete, or a greased, tapering wooden plug to be drawn out when concrete has set.

A trough with a solid concrete base should be made in the same general way, the forms carried up to the desired height of trough, and the reinforcing embedded in the concrete a few inches from the inside. Troughs should be protected from the sun and currents of air for several days, and kept wet by sprinkling.

The method of construction of a small concrete tank is plainly shown in the accompanying illustration (Fig. 25). A tank of this size is suitable for ordinary purposes. See also Plate 6.

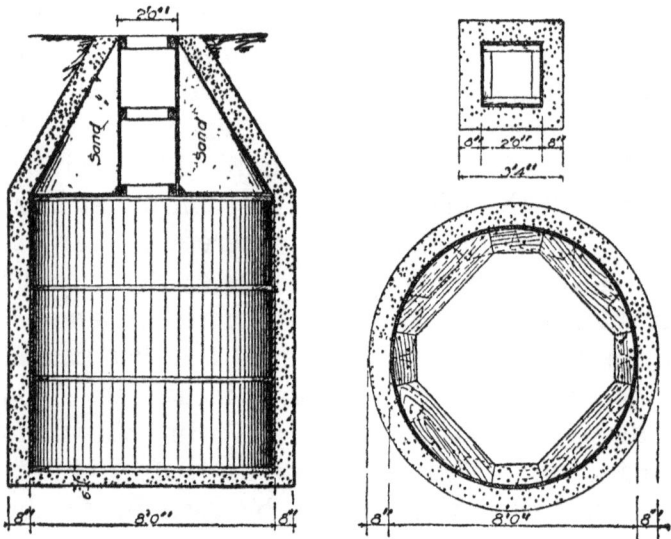

Fig. 26. Details of Construction of Concrete Cistern.

A Concrete Cistern. The mixture for a cistern should be made of one part cement and three parts sharp sand. The two should be mixed dry, and water sufficient to make a stiff mortar should be added. If the soil is of clay or of such a character that there will be no danger of a cave-in, it is not necessary to wall up with brick. The mortar should be applied about one-half inch thick, and should be followed immediately with a second coat one-fourth of an inch thick. Then give a skim coat made of equal parts of cement and sand. Keep the surface moist a week, and do not fill with water until 10 or 12 days. See Plate 7.

Cisterns made in this manner are usually built egg-shaped, with the use of falsework for crowning the top, in which case the skim coat cannot be added. The walls in egg-shaped cisterns should be two inches thick at the bottom, three-quarters of an inch on the side, and three inches on the crown. If there is to be more than the ordinary pressure on the crown, it can be reinforced with wire netting. Local conditions must favor this method; it will not answer in all cases. A four-inch brick wall laid in cement mortar (one part cement and two parts sand), and plastered on the inside a half-inch thick with the same mortar, waterproofed, is in most cases satisfactory. But to make the cistern absolutely water-tight, let the wall become hard, which requires six or eight days.

For a round cistern of large dimensions, make a circular excavation 16 inches wider than the desired diameter of the cistern, or allow for a wall two-thirds the thickness of a brick wall that would be used for the same purpose, and from 14 feet to 16 feet deep. Make a cylindrical inner form the outside diameter of which shall be the diameter of the cistern. The form should be about 9 feet long for a 14-foot hole, and 11 feet long for one 16 feet deep. Saw the form lengthwise into equal parts for convenience in handling. Lower the sections into the cistern and there unite them to form a circle, blocking up at intervals six inches above the bottom of excavation. Withdraw blocking after filling in spaces between with concrete; and then fill holes left by blocking, with rich mortar.

Make concrete of one part Portland cement, two parts clean, coarse sand, and four parts gravel or broken stone. Mix just soft enough to pour. Fill in space between the form and the earth with concrete, and puddle it to prevent the formation of stone pockets, using a long scantling for the purpose, and also a long-handled paddle for working between the concrete and the form.

To construct the dome without using an expensive form, proceed as follows: Across the top of the form build a floor, leaving a hole in the center two feet square. Brace this floor well with wooden posts resting on the bottom of the cistern. Around the edges of the hole, and resting on the floor described, construct a vertical form extending up to the level of the ground.

Build a cone-shaped mould of very fine wet sand, from the outer edge of the flooring to the top of the form around the square hole, and smooth with wooden float. Place a layer of concrete four inches thick over the sand, so that the edge will rest on the side wall.

Let the concrete set for a week; then remove one of the floor boards, and let the sand fall gradually to the bottom of the cistern. When all boards and forms are removed, they can easily be passed through the two-foot square aperture, and the sand taken out of the cistern by means of a pail lowered with a rope. This does away with all expensive forms, and is perfectly feasible. The bottom of the cistern should be built at the same time as the side walls, and should be of the same mixture, six inches thick.

A **square cistern** is much easier to build, and in most cases answers the purpose as well as a round cistern.

Excavate to desired depth, and put in 6 inches concrete floor, one part Portland cement, two parts sand, and four parts broken stone. As soon as practicable, put up forms for 8-inch walls, and build the four walls simultaneously. If more than 8 feet square, walls should be reinforced with a woven wire fabric or steel rods.

Concrete Well Curbs. For cleanliness and sanitation, so essential in a well, concrete possesses advantages over brick or stone, especially if the surface down to the water line is made waterproof to keep out any possible seepage of surface water. A method that is recommended for the building of well curbs is here given that is applicable to almost all cases. The excavation having been carried down to water, the sides of the well made smooth and made ready for the concrete, a form ten inches less in diameter than the well excavation should be made of planks nailed securely and vertically to the frame. This form should be at least two feet in height, and may be higher, but a two-foot form will be handled more easily.

Operations should commence at the bottom of the well, where the form is first placed. The mixture for the concrete should consist of one part Portland cement, three parts coarse and sharp sand, and five parts of gravel or broken stone. A waterproofing also should be used. Placing the form in the center so that a five-inch margin will remain around the circumference for the concrete, begin filling in, tamping every four inches. When the form is filled nearly to the top, allow the concrete to set, and raise the form carefully for the next section, being careful to have it at all times plumb. This operation is repeated with the filling in till the top of the well in reached. The concrete should be laid five or six inches above the surface of the ground to keep out all surface water. As the work of filling in progresses, staging of planks can be built to provide working room. After the concrete is dried, these can be taken out.

HOLLOW-WALL CONCRETE SILOS

Throughout the northern section of the United States and in Canada, the introduction of the hollow wall—continuous dead air space—in concrete silo construction has met with general favor and approval. The advantage of this form of silo is apparent to those living in a country exposed to very low temperature for prolonged periods of time. With proper attention to the closing of doors, silage can be kept through the coldest winter without freezing. While the question of the use of frozen silage is still being argued by our able State Experimental Stations, it is certain that although freezing may not hurt the silage and may not be injurious to the animals that feed on it, it certainly does not add anything of value to the silage. This, in connection with the trouble of thawing it out before feeding, makes the use of a silo that is proof against freezing, of a decided advantage. Wood silos with double walls have been built throughout the Northern States in an attempt to prevent this freezing; but while they have succeeded fairly well in this feature, have in a comparatively short time failed by the rotting of the wood. The silage juices get into the dead air space and rot out the wood in a remarkably short time. Careful ventilation during the warm weather may lengthen the life of the wood for possibly a year or two, but even this ventilation will not prevent rotting for a longer time. The only advantage claimed for the best double-wall wood silo—lined with cement plaster—has been its greater cheapness. With the reduction in cost of the concrete silo, this advantage has now practically disappeared.

This reduction in the cost of the hollow-wall type is due to the use of reinforcement and of specially adapted forms for moulding the green concrete. There are a number of companies operating who build the silo for the farmer, or furnish him with the apparatus or plant to build it himself, with full instructions as to how to build with his own labor. This saves considerable money on the item of labor alone.

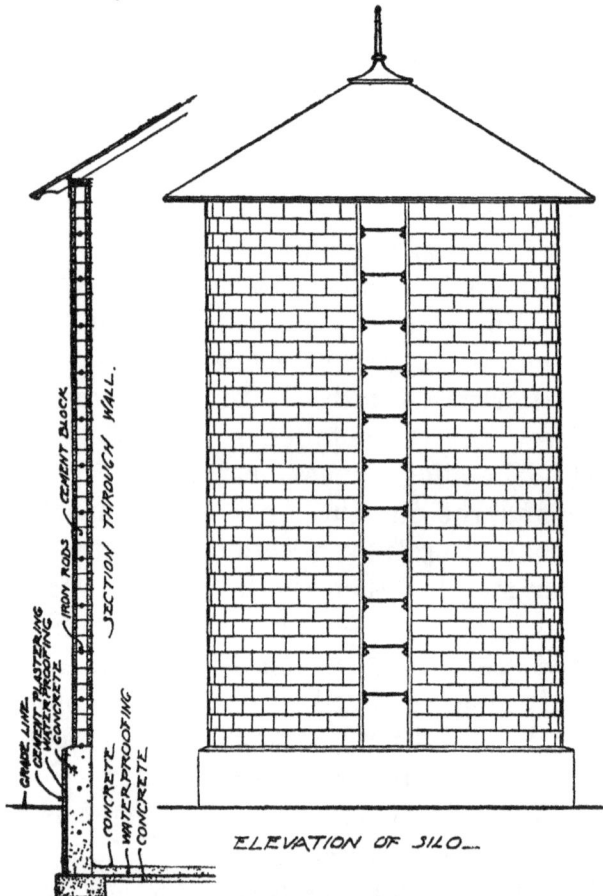

Fig. 27. A Concrete Block Silo.

Advantages of Concrete Silos. Fire is the farmer's greatest dread. When a fire starts from lightning or any other cause, the farm buildings usually burn down. For this reason, insurance rates are very high, and farmers find it a great tax to protect themselves by carrying insurance. The buildings, however, are not the most serious loss; fires most frequently occur during the latter part of the summer or in the early fall, after the crops have been harvested; and, although the buildings can be replaced, practically the year's work of the farmer is gone. Very often the fire spreads so rapidly that the stock is also lost.

A concrete silo cannot burn down, as concrete is fireproof; nor can the food stored in it be either injured or destroyed. A temporary structure can be erected to replace the burned building, but the crops cannot be replaced, except at great expense.

Insurance companies have recognized the indestructible qualities of concrete by making an insurance rate so low as to be within the means of every farmer.

The only objection that has ever been made to a concrete silo is its cost. The cost varies, owing to the price of materials of which the concrete is made. In many places concrete silos are cheaper than any other kind; few farmers are without a gravel pit suitable to furnish both gravel and sand of a quality proper for making good concrete. Moreover, Portland cement can now be obtained at a reasonable cost. Under these conditions, a concrete silo is cheaper than any other kind.

The best is the cheapest, regardless of the original cost. A silo which never leaks, will not blow over, is always ready to be filled without first repairing, requires no repairs, cannot burn down, and is vermin-proof, is certainly the best and cheapest. A concrete silo is all of these.

Kinds of Concrete Silos. Three kinds of concrete silos have been in successful use for several years. These are known as **monolithic solid-wall silos, monolithic hollow-wall silos, and concrete block silos.** All three are good; and in choosing between them, the cost, which is fixed by local conditions, should be the deciding feature, unless the location of the farm is so far north that the extreme cold in winter makes a hollow monolithic wall or hollow block wall silo preferable to prevent the freezing of the silage. See Plates 8 and 9.

Size of the Silo. The size of the silo depends upon the **number of cattle to be fed** and on the **number of days their feeding continues.** It does not pay to build a silo for less than ten head; but, as someone very aptly put it, "Build a silo and get the ten head to keep."

The diameter, inside measurement, should never be more than one-half the height, and in practice it is not found advisable to make it over 20 feet.

For convenience, Table III has been pre-

TABLE III

Dimensions and Capacities of Silos

NUMBER OF COWS IN HERD	FEED FOR 180 DAYS				FEED FOR 240 DAYS			
	ESTIMATED TONNAGE OF SILAGE CONSUMED	SIZE OF SILO		CORN ACREAGE REQUIRED AT 15 TONS TO ACRE	ESTIMATED TONNAGE OF SILAGE CONSUMED	SIZE OF SILO		CORN ACREAGE REQUIRED AT 15 TONS TO ACRE
		DIAMETER	HEIGHT			DIAMETER	HEIGHT	
	TONS	FEET	FEET	ACRES	TONS	FEET	FEET	ACRES
10	36	10	25	2½	48	10	31	3½
12	43	10	28	3	57	10	35	4
15	54	11	29	4	72	11	36	5
20	72	12	32	5	96	12	39	6½
25	90	13	33	6	120	13	40	8
30	108	14	34	7½	144	15	37	10
35	126	15	34	8½	168	16	38	11
40	144	16	35	10	192	17	39	13
45	162	16	37	11	216	18	39	14½
50	180	17	37	12	240	19	39	16
60	216	18	39	14½	288	20	40	19
70	252	19	40	17	336	—	—	—

pared, showing the size of silo required for feeding any number of cattle for a given time.

This table gives the number of cows in herd, and tonnage of silage, for both 180 and 240 days of feeding of 40 pounds of silage per cow; also acreage of corn estimated to fill the silo, and the dimensions of the silo itself. The diameters given are such that at least 2 inches in depth of silage will be taken off daily.

As stated above, the number of animals to be fed should determine the diameter of the silo, and the length of time silage is wanted should determine the height of the silo. The amount of silage to be fed per cow must be determined first. Decide whether each cow is to have 20, 30, 40, or 60 pounds per day. Then, having decided this point, make the diameter of the silo such that by feeding the cows so much per day the silage can be fed down at least 2 inches per

TABLE IV

Approximate Capacities of Round Silos

(The diameter is shown at the top of the columns, and depth at the left)

INSIDE DIAMETER OF SILO IN FEET, AND CAPACITY IN TONS (2,000 LBS.)

HEIGHT OF SILO	10 ft.	11 ft.	12 ft.	13 ft	14 ft.	15 ft.	16 ft.	17 ft.	18 ft.	19 ft.	20 ft.
FEET	TONS	TONS	TONS	TONS	TONS	TONS	TONS	TONS	TONS	TONS	TONS
20	26										
21	28										
22	30	36									
23	32	39									
24	34	41	49								
25	36	43	52								
26	38	46	55	64							
27	40	49	58	68							
28	42	51	61	71	83						
29	44	54	64	75	87						
30	47	56	67	79	91	105					
31	49	59	70	83	96	110					
32	51	62	74	86	100	115	131				
33	53	65	77	90	105	121	138				
34	56	68	80	94	109	126	143	162			
35	58	70	84	98	114	132	149	169			
36	61	73	87	102	118	136	155	176	196		
37	63	76	90	106	123	142	161	183	204		
38	66	79	94	110	128	148	167	191	212	237	
39	68	82	97	115	133	154	174	198	221	247	
40	70	85	101	119	138	160	180	205	229	256	280

day, as this will prevent moulding of silage. If a silo is made too large in diameter, and this is the most frequent error, one of two things will happen—First, the silage will be mouldy all the time, owing to the inability to feed it down rapidly enough; or, second, the cows will be fed more than they should have in an attempt to keep ahead of the moulding.

Where large cows are kept, and it is expected to feed 40 or 60 pounds per cow daily, it frequently happens that it is desirable to cut down the silage ration. It is well to have the diameter of the silo small enough so that the farmer can cut down the ration one-third or even one-half.

and still be able to feed down the silage 1 to 1½ inches daily.

In the dairying sections, many farmers consider this point so important that they are building two small silos instead of one large one, so that they can feed a light ration and still feed down the silage rapidly enough to prevent moulding. In the older dairying sections where silos have been longest in use and where dairymen have used up their first silo and are building a second time, they build two small ones in place of the one large one. They build smaller in diameter and higher.

About fifty cows seems to be the most that can be fed with advantage from one silo. In general, 40 pounds of silage is figured as the average daily feed of a cow.

For further convenience in figuring in connection with the sizes and capacities of silos, Table IV is presented.

Silo of Concrete Blocks. The method used for the construction of a silo of this character is as follows: Place five eight-inch rods with turnbuckles every four or five feet, the ends being down in the hollow block sufficiently to hold while tightening the turnbuckle. The lap or tie rods used around the entire circle at the same course of blocks, are made of half-inch rods. The jamb at the opening is made of 2 by 12-inch wood, with a 1 by 4-inch board setting into the recess of the block to prevent slipping, and two stops on the opening side for a sliding plank. A casing board on the inside is required only when the blocks have no recess. It may prove cheaper to build a silo of 2 by 6-inch studding, 16-inch centers, with metal lath and stucco plaster on both side and interior.

Construction of a Concrete Silo. The silo has come to be one of the necessary parts of the farm equipment, and it is unquestioned that concrete is the ideal material for its construction, for concrete makes the silo air-tight and too heavy to blow over, and it will last for ages. The concrete silo can be built at low cost, provided the walls are not made too thick. By using reinforced concrete, the thickness of the walls can be reduced to a minimum. Silos in cold climates are best built with a hollow wall. The specifications for one of this character are:

Excavate to a depth below the frost line and of the desired diameter, allowing for the thickness of the walls. Erect a sixteen-inch solid wall to the level of the ground with concrete, one part Portland cement, two and one-half parts clean, coarse sand, and five parts broken stone. After removing forms, fill the excavation inside the walls to within 8 inches of the ground level, with cinders, gravel, or broken stone, and tamp hard. Pick with a stone ax that part of the inside wall that shows above the porous foundation, and wet thoroughly. Fill the space on top of the cinders, etc., with concrete, to within one inch of the foundation.

Erect forms four feet high for three-inch hollow-core walls with ten-inch air-chamber.

In a circular form there are two sides, the inner and the outer. These are made in the same way, but cannot be of the same pattern, as the thickness of the walls comes between the two parts, making the radius of the sides different. The simplest way to make a circular form is to draw a circle of the size of the form desired and lay boards around the circumference of the circle. These boards should be lightly tacked together in place, and, using the same measure, mark the circle on them. They should then be knocked apart and sawed out along the lines marked, the pieces then being fastened securely together. After making two or more circular forms, place them at equal distances apart, and put on the side boards.

CONCRETE IN BARN AND STABLE CONSTRUCTION

Barn Foundations. These are laid in the same manner and of the same proportions in the mixture as house foundations, except that there is no cellar unless it is desired.

Barn and Stable Floors. The general rules of sidewalk construction, given below under the head of "Sidewalk Construction," apply to barn and stable floors. The thickness of the porous sub-base for a **barn floor** should be 6 inches to 12 inches, the base 3 inches to 5 inches, finishing with a surface of mortar, one part Portland cement and one and one-half parts clean, coarse sand, 1 inch to 1½ inches thick. This may be roughed at time of laying and before it has set, or grooved in blocks about 6 inches square, to prevent the animals slipping. The surface should have sufficient slope to carry liquids to drains placed at convenient intervals. These drains may be either gutters or pipes laid under the floor, leading to a manure pit. If pipes are used, they should be laid in the sub-base, and the joints put together with cement mortar, care being taken to give the pipes enough slope to flush properly, and making them of straight lengths between openings so that they can be cleaned if necessary. The lids of the drain

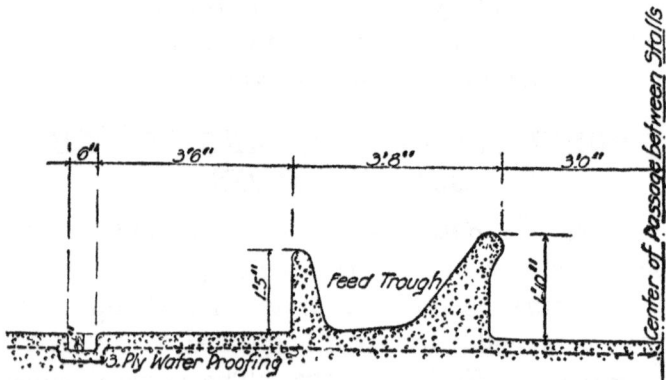

Fig. 28.   Section Showing Details of Concrete Floor for a Cow Stable.

should be sunk about ¼ inch below the level of the floor, and should be loose, so that they can be removed conveniently.

Several years' experience in the use of concrete for barn floors and drains proves that

A CONCRETE ICE-HOUSE.

A CONCRETE CISTERN.

PLATE 7—CEMENT CONSTRUCTION.

SILO WITH CONTINUOUS HOLLOW CONCRETE WALLS.

Inside diameter, 16 ft.; outside, 18 ft.; dead air space, 3½ in.; inside wall, 5½ in. thick, reinforced with fence wire; outer wall, 3 in.; height, 34½ ft. Built for Illinois Farmers' Institute in 1905.

CONCRETE BLOCK SILO.

Blocks 8" by 10" by 16", with two air chambers; tied together with steel ties and wire binders between each block. Height of silo, 29 ft.; diameter, 15 ft.

PLATE 8—CEMENT CONSTRUCTION.

manure will not injure well-made concrete, provided the concrete has thoroughly set and hardened before use.

Driveways are made by dividing into 6-inch squares to give foothold.

The dairyman and agriculturist are more and more coming to recognize the concrete floor as the ideal for barn and stable. Excavation for a **stable floor** should be made below the frost line, and there should be a sub-foundation of at least six inches, and even more if possible, depending upon the weight and wear the floor is to have. A deposit of five inches of concrete should be made upon this sub-foundation, consisting of one part Portland cement, three parts sand, and five parts gravel or crushed stone, well mixed. The top coat should be two inches thick, one part cement and two of sand. The surface should be so sloped that the liquid manure and water of the stable will flow to some desired point for drainage away. The top should be grooved before it sets, to give the animals foothold and prevent their slipping. If the floor is to be of more than ordinary size, it should be laid in sections, and provided either with sand joints or the sections separated by pieces of tar paper.

Feeding Floors. The immense advantage of concrete feeding floors over the old method of placing fodder on the ground, is apparent to all who have given the subject any thought. Feeding floors should be built the same as sidewalks.

The finishing coat is optional, although it has the advantage of being much easier to keep clean. Many farmers prefer an unfinished surface, on account of its giving cattle a firmer footing in slippery weather.

Box Stalls. There is nothing so warm in winter or cool in summer as a concrete structure. Concrete box stalls are of immense advantage on this account, as they prevent a horse becoming restive and ill-tempered. They may be built of concrete one part Portland cement, two and one-half parts clean, coarse sand, and five parts broken stone. The walls should be 4 inches thick, and reinforced with one-quarter-inch steel rods 12 inches apart. A smooth surface can be secured by plastering the walls one-eighth inch thick with mortar, one part Portland cement and one part clean, fine sand, after they have been picked with a stone ax and thoroughly wet.

A concrete water box and manger may be built in with the same mixture as the mortar used in plastering. See Plate 1 (lower figure); also Plates 10 and 11.

Grooved and Roughened Surfaces. The stable yard and the stable floor, carriage runways, and all places where it is intended to have animals use the concrete floor, are made better by being grooved or roughened. This treatment of concrete prevents the animal from slipping, and gives a foothold to the floor, besides affording drainage for water to some common point where it can find egress. Concrete surfaces may

be roughened also by beating them with a brush
before they are hardened.

Piers and Posts. For this work the builder
should excavate below the frost line, and build
forms 2 feet square to a point within 6 inches
of the surface of the ground. Fill with concrete
consisting of one part cement, two and a-half

Fig. 29. Use of Forms in Construction of Piers and Posts.

parts clean, sharp sand, and five parts gravel or
crushed stone not more than one inch in size.

The mixture must be tamped carefully. From the center of this foundation, build a hollow form one foot square and to the desired height, and fill with concrete of the same composition as the other. Before the form is filled, and in fact before it is set, place four steel bars $\frac{3}{4}$ inch in diameter, vertically, so that they will be about two inches inside the corners; and around them, at intervals of one foot, wind loops of $\frac{1}{8}$- or $\frac{1}{4}$-inch wire, tying them to the steel rods with finer wire. Every two feet a short piece of $\frac{1}{2}$- or $\frac{1}{3}$-inch wire may be tied to each of the vertical rods so as to project against the form and hold the steel in place. The concrete should be made soft and pliable so that it will flow, and, as it is poured into the top of the mould, work a long paddle, made like the oar of a rowboat, against the forms, to force the stones away from the surface and drive out bubbles of air which tend to adhere to the forms and form pockets. This method of construction makes an excellent foundation for a barn.

A Concrete Rain Barrel. Among the newer uses to which concrete has been placed is that of making rain barrels. A convenient size, easily made, is 36 inches high, 24 inches in diameter, with a shell $\frac{1}{2}$ inch thick, the whole being treated with a waterproofing compound. No matter how long such a barrel is left empty in the hot sun, it will not spring a leak.

A Horse Block. In the construction of a horse block, the method does not differ materially from that employed in making a small tank. Build a box 24 inches long, 10 inches wide, and 8 inches deep, outside measure. Turn this bottom up on the floor or some other smooth surface; and around it build a box or form, without bottom, 36 inches long, 18 inches wide, and 12 inches deep, inside measure. Be sure that the smaller box is set at equal distances from both sides and ends of the larger box, and fill the form thus made with concrete, one part Portland cement, three parts clean, coarse sand, and five parts gravel or broken stone. Scrape with straight-edge, and smooth with wooden float. Let it stand for at least 48 hours before removing outside form. Keep damp by sprinkling for three weeks, and do not attempt to move it before that time. If finished appearance is desired, as soon as the form is removed a coating one-eighth inch thick, made of one part Portland cement and one part clean sand may be plastered over the entire surface of the block, after picking with a stone axe and wetting thoroughly. See Plate 6.

Hot-Bed Frames. Excavate a trench to a depth below frost, and erect forms for a 4-inch wall. Fill with concrete mixture one part Portland cement, three parts clean, coarse sand, and six parts gravel or broken stone that will pass a half-inch sieve, to level of the ground. On top of these, build forms for a 3-inch wall to the height desired, and fill with concrete of the same proportions. These structures are so small in size that no reinforcement is necessary in the walls. On the upper edge of the walls, and around the interior, may be embedded strips for use when the glass frames are placed on top. Remove the forms in two or three days, and keep the walls damp for a couple of weeks.

Greenhouses. The concrete greenhouse offers the special advantage of being more easily heated than a wooden one. Greenhouses of this construction also keep out the cold air, and protect the growing plants against sudden changes of temperature.

The greenhouse foundation should be ten inches wide and sixteen inches deep. The mixture should be one part Portland cement, three parts sand, and six parts crushed stone or gravel. On this, and at equal distance from each edge, a wall seven inches thick should be built. The mixture for this should be one part Portland cement, two and one-half parts sand, and five parts cinders. The wall should be carried up to the height desired, and a ridge-pole erected six inches wide and eight inches deep, with one part Portland cement, two and one-half parts sand, and five parts crushed stone or gravel not more than three-quarters of an inch in size. This pole should be reinforced with two steel bars half an inch in diameter. If the total width of the house is not over 16 feet, beams from the ridge-pole to the side walls 2½ inches by 5 inches, reinforced with a half-inch bar, will be strong enough. For the support of the ridge-pole, posts eight inches square should be placed at intervals of ten feet. The final or finishing dress for all the concrete should be a quarter-inch coat of cement mortar.

A Concrete Fountain. In the adornment of the rural lawn, nothing adds so much to the attractions of a home as a fountain. Any family owning a windmill and elevated tank has the equipment for an improvement of this kind that will be a source of delight in the summer months. The plan here suggested is for a fountain having three basins and six feet high. The lower or fountain basin, whose edges should be elevated above ground-level about four inches, is five feet in diameter. The middle basin is three feet, and the top two feet in diameter.

The first step, after the desired size is determined, is to make the pipe connections and provide for the feeding arrangements for the fountain. A pipe frame or skeleton of pipe should be made, with the provisions for the spouts of water, all of which should tend a little inward to get a centralization of the streams. The pipe should be of 3/4-inch size to assure a free flow, except the central or "trunk" pipe, which should be 2 inches. Four pipes, quartering the circle, should be joined to the center or vertical feed pipe.

The excavation for the ground basin should be made below the frost line, and provision made for draining the fountain by laying pipes that will carry the water off through an overflow vent. A foundation of crushed stone or gravel should be laid; and after this has been watered, the concrete can be laid, leaving a hole in the center for the pipe. A circle may be drawn with the pipe-hole as a center, and the work of laying the concrete should be carried on with this as the working point. The concrete, one part Portland cement, three of sand, and five of gravel, should be well tamped.

When the concrete has dried sufficiently to allow the work to proceed, attach the pipe system, and place a circular form for the first section of the central column, tapering up slightly and about 12 inches in diameter at the bottom. Fill with concrete, and tamp in layers of four inches. Place a saucer-shaped form far enough below the pipes of the middle basin so that they will be in the center when the concrete is filled in. This form should be about two inches more in diameter than the circle of pipe. Fill with concrete, trowel thoroughly. and proceed with the second section of the central column and top basin in the same manner.

The forms should not be removed for at least two weeks, at the end of which time they may be taken away, and the fountain given a finishing coat of waterproofed mortar.

The forms for the basins will be best made of 2 by 6-inch lumber, each piece sawed out so as to give an approximate saucer shape when all are nailed together. Then the finish of the form may be done with a chisel. Ingenuity will find it possible to give the edges of the basins artistic effects in the finishing.

ROAD CULVERTS

It may be said of the concrete culvert that it will be found intact after a flood or freshet, while brick or stone culverts in the same neighborhood will be washed away. It is true, too, that they are cheaper than culverts made of any other material, besides being more durable. It need only be mentioned that the only time to build a culvert is in a dry season when there will be no difficulty with water. When this is not practicable, the water should be diverted, if possible.

In this work as in all other operations where concrete is to be placed on the ground, excavation must be carried below the frost line. Trenches for the foundation should be dug on each side of the bed of the stream. The concrete should be of one part Portland cement, three parts sharp, clean sand, and six parts of gravel or crushed stone. Build an apron with this concrete across the bed of the stream between the two foundations, and with its level equal to the bed of the stream. For a culvert of ordinary length this should be eight inches thick. Place the semi-circular arches of the size required, and brace them well. Two-inch plank should be used for the forms. After the forms have been well greased, fill in the concrete, tamping thoroughly every four inches. A reinforcement of expanded metal should be placed inside the surface of the arch about two inches.

Full, detailed practical information on the subject of the construction of the larger types of reinforced concrete culverts such as are used in railroad work, will be found elsewhere under the head of "Reinforced Concrete Culverts."

USE OF COLLAPSIBLE FORMS

By actual use concrete culverts have demonstrated their economy. A concrete culvert can be constructed by the farmer, and will be more serviceable than one made of any other material. Its first cost is low and there is practically no further expense. No repairs are necessary. Properly constructed, freezing does not affect it, thawing does not damage it, and it becomes harder and harder every year. It solves the problem of an unobstructed bore, for no joints are necessary, and there are no projections to catch obstructions that might clog a clay tile or corrugated culvert. We have already, in connection with the subject of "Forms," referred briefly to several types of collapsible metal forms and the method of using them in the construction of small culverts.

The accompanying table gives figures from which to make estimates, giving the necessary thickness of top, bottom, and side walls for various-sized culverts, the amount of material needed, etc. These figures are for culverts eighteen feet long. On longer or shorter ones,

TABLE V

Data for Estimating on Culvert Construction

Diameter of Culvert Inches	Waterway Squared Inches	Thickness Bottom Inches	Thickness Sides Inches	Thickness Top Inches	Cubic Feet Sand Required	Pounds Cement Required	Width To Dig Ditch Inches	Complete Cost
8	6x8 37	3	4	4½	30	465	16	$ 5.40
11	10x9 50	3	4	5	36	558	19	6.84
13	12x11.60	3	5	6	48	744	23	8.64
15	12x14 75	4	5	6	59	915	25	10.62
19	16x17.72	4½	5½	6½	85	1317	30	15.30
22	19x20	4½	5½	6½	90	1395	33	16.20
27	24x23.75	5	6	7	124	1922	39	22.32
38	24x41	5	6	7	165	2557	48	29.70
48	36x57 50	6	7	8	255	3952	62	49.90

make your estimates in proportion. In the table you will find the squared dimensions of waterway for which different sizes of cylindrical moulds, such as the "Overturf," will be the equivalent; also the thickness of the sides, bottom and top walls, the width to dig the trench, and the amount of sand or gravel, and of cement to be used.

The costs of completed culverts as here estimated are figured on what are considered average conditions, with cement at 65 cents per hundred pounds, and gravel at $1 per yard. The cost of this kind of concrete work is close to 18 cents per cubic foot of concrete used; and, as cement in the proportions mentioned does not increase the bulk of gravel, to estimate the cost of any sized culvert completed, multiply the estimated amount of gravel in cubic feet by 18—this will give you the cost complete. Multiply by 11 to get the cost of cement; by 7 for cost of material on the ground and placed.

The figures in the table are made on a basis

of 18-foot culverts, with coping to extend two feet from waterway through culvert, and concrete in proportion of one part cement and six parts gravel or sand.

If wooden forms are to be used, excavate trenches for foundation to a depth below frost and 2 feet 8 inches wide; and at the upper end of the culvert, connect the two foundations across with an 8-inch wall the height of invert. This is called an **apron,** and will prevent scouring. Build invert 8 inches thick, having the top on a level with the bed of the stream. Next build forms for the wall, with one straight form strong enough to support the arch, and well braced, and the other form as the thickness of the wall requires.

For convenience in keeping the road open for traffic, and the saving in material for forms, we suggest making only nine feet of the culvert at a time. Should this suggestion be accepted, proceed as follows:

Make three semicircular forms the size required, out of $1\frac{1}{2}$-inch stuff, and set them in place three feet apart. Fasten joist 2 inches by 4 inches by 9 feet on them. This is called **lagging.** Set the form thus made on large wedges supported by top of form marked "Sill." Grease forms well, and fill with concrete of a rather wet consistency, and tamp thoroughly every 6 inches, taking care not to disturb the form. Let stand until thoroughly dry, about 28 days; and then knock out the wedges, lowering the semicircular form, which will be easy to remove.

Should the culvert be made all at one time, enough semicircular forms should be constructed to support the lagging at least every 3 feet.

Reinforce the concrete with expanded metal, placing it so that it is 2½ inches in from the under side of the arch, and extending down through the walls. All concrete should be mixed one part Portland cement, three parts sand, six parts broken stone. Should wing-walls be required, they should be built at the same time as the foundation, should go to the same depth, and be reinforced, the reinforcing connecting with that in the walls. The width of these walls should be left to the judgment of the man in charge of the work.

CONCRETE SEWER PIPE

In the draining of marsh land and for sewerage purposes, pipe made of concrete—if properly made—is not surpassed by any other material. Pipe of this kind has been made for many years in Germany, the matrix being Portland cement. Rudolph Hering, a New York engineer, who has investigated the subject, says:

"Cement sewer pipe has a competitor in the United States perhaps to a greater extent than elsewhere, in vitrified clay pipe, which is very extensively made in our country, and is still almost exclusively used for small sewers. As cement pipe can be made cheaper than clay pipe, it is naturally forcing itself into use.''

Mr. Hering claims the following advantages for concrete sewer pipe over the clay variety:

A sectional form can be given them which is more conducive to stability and efficiency than the round form customary in clay pipes.

As vitrified pipe warp in burning, the section is not finished truly circular, and slight projections are formed at every joint when the pipes are laid to form a sewer.

As cement pipes have a truer sectional shape than vitrified pipes, they can be given a slanting butt joint, as is customary in Europe, instead of the more costly bell and spigot joint common for vitrified pipe, which are made in imitation of cast-iron pipe used under high pressures.

Concrete pipes are tougher and less brittle than vitrified pipes.

Concrete pipes, if well made of proper materials, have a strength to resist compressive, tensile, and bursting strains which is amply sufficient for all purposes for a sewer in a large city. If the materials are carefully selected, the concrete pipe should be as permanent as the vitrified pipe. Concrete work in the sewers of Paris several hundred years old is as sound to-day as when it was laid.

For the manufacture of concrete sewer pipes, a number of machines are on the market; and when the aggregate used is carefully selected and cleaned, the mixture being of 1 part Portland cement, 2½ of sharp, coarse sand, and 4 parts crushed stone not over half an inch in size, a good pipe should be turned out.

Circular forms of steel should be used, and the tamping should be done in the same manner as for concrete blocks.

"Mercantile" Pipe. A type of reinforced concrete sewer pipe adapted for farm building drainage purposes is that known as the "Mercantile." It is reinforced with wire, and has a flanged interlocking joint. After the pipe is laid, the exposed part of the joint is filled with concrete, making a practically monolithic pipe.

A CONCRETE CHICKEN HOUSE

It is easier to keep a chicken house clean when it is made of cement than it is to care for such a house built of any other kind of material. Besides, the concrete chicken house is rat-proof. The protection against rats, weasels, etc., and the ease with which such a structure is kept clean, should be sufficient reason to give it preference over every other kind.

Excavate a trench 12 inches wide, to a depth below frost, and fill with concrete one part Portland cement, three parts clean, coarse sand, and six parts cinders. On this foundation, and at equal distance from either edge, build a solid wall 5 inches thick, one part Portland cement, two and one-half parts clean, coarse sand, and five parts cinders; or, if cinders are not obtainable, a hollow wall should be built 12 inches thick, consisting of two 3-inch walls and a 6-inch air space. The roof may be made of wood or of concrete. If the house is not more than 8 feet wide, a roof with slope in one direction may be made of a 4-inch concrete slab reinforced with steel rods or heavy wire mesh. For a shorter span, a less thickness may be adopted.

Hens' nests are also made of concrete. They are vermin-proof, and are adapted to maintaining the desired evenness of temperature. They can be washed out, or kept in perfect sanitary condition by filling with straw or other combustible material, and burning out.

Poultry House of Concrete Blocks. The following is a plan for a poultry house. The front is 6 feet high and the rear 4½ feet. The ground space is 8 by 16 feet. A partition is built 6 feet from the front, leaving an 8 by 10-foot scratching shed. This shed has an open front 3 feet high, with a wire netting stretched across, and a curtain to drop in rough weather. The roof is made of cement mortar, but not in the ordinary way. The rafters are 2 by 4-inch, set to 16-inch centers. The roof is lathed with strips 1 by 2 inches set ¼ of an inch apart.

The mixture for the roof is the same as for cement mortar—one part Portland cement and two parts sand. Strips should be nailed around the edge of the roof, extending five-eighths of an inch above the edge as a straight-edge for the finishing. The cement should be troweled down

CONCRETE BLOCK SILO.

Height, 38 ft.; inside diameter, 19 ft. 6 in.; blocks, 8 in. thick, with slots in ends, which after laying were filled with concrete, thus giving bond.

SILO BUILT OF CONCRETE BLOCKS.

Height, 35 ft.; inside diameter, 14 ft.; blocks in lower third 8" by 10" by 16"; in upper part, 8" by 8" by 16". Every third course reinforced with ½" round iron embedded in grooves in blocks.

PLATE 9—CEMENT CONSTRUCTION.

smooth. The roof made in this manner will prove eminently satisfactory in every case.

Another plan provides that a trench 12 inches wide should be dug below the frost line. Fill it with concrete consisting of one part cement, three of sand, and six parts gravel or crushed stone. The walls of the chicken house, which should be reinforced with ½-inch iron rods a foot apart, should be five inches thick. For these, forms will be necessary, allowances being made for the windows, which should be on the south side of the little structure. The roof may be made with one slat, and of reinforced slabs of a 1:3:5 mixture, with wire netting reinforcement. The concrete for the walls should be of one part cement, three parts sand, and five parts gravel or crushed stone that will pass a half-inch mesh.

A concrete poultry house will be found to prevent lice in chickens more readily than a wooden structure that has to be whitewashed frequently, besides being warmer in the winter and thus encouraging hens to lay when eggs are at their highest.

HOG PENS OF CONCRETE

Usually, on the farm, the greatest difficulty in the matter of cleanliness about the buildings is encountered in connection with the care of the hog pens. By the use of concrete, whose surface lends itself readily to flushing with water, much of this trouble is avoided. The size and shape of the pen having been decided upon, a trench for the foundation should be excavated a foot wide and below the frost line. In this, a foundation consisting of one part Portland cement, three parts sand, and six parts gravel or crushed stone, should be laid. Four parts of gravel or crushed stone is used in the mixture for the wall, instead of six parts as in the foundation. The floor of the pen is made in the same manner as followed in the laying of concrete walks.

The hog house may be built in one corner of the pen, and its walls should be four inches thick, with a reinforced one-slant roof of 2½-inch concrete reinforced with netting and with rods of ½-inch thickness placed 10 inches apart, if the house is not more than 10 by 12 feet.

A trough can be made by the use of two forms, one of a long box shape, and the other shaped like a V for the trough cast. First pro-

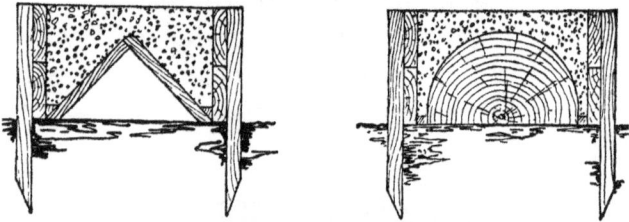

Fig. 30. Forms for Hog Troughs.

vide a smooth surface or platform to lay the forms on. Place first the V form in the position of a pyramid. Over it set the box, being careful that the adjustment is arranged so the V form is in the middle of the box. Fill with a mixture consisting of one part Portland cement, three parts sand, and five parts gravel or crushed stone. Do not disturb the forms for three or four weeks.

A CONCRETE ICE-HOUSE

The modern equipment of a rural home, especially if it is located where a supply of ice is available for harvest in the winter, is not complete without an ice-house. And for this purpose no material lends itself better to the construction than does concrete. To be sure that the heat will be kept out in the summer, it is well to provide for a wall at least sixteen inches in thickness. A wall of this character will protect the contents of the ice-house amply. A hollow wall, also, is advantageous.

For an ice-house of ordinary size, sufficient to hold a supply for a family of five or six, an excavation one foot below the desired depth of foundation will do. This surface should be cleaned off and leveled, and upon it a layer of crushed stone or broken bricks should be placed, ramming the material thoroughly. This will afford opportunity for drainage. See Plate 7.

If the wall is to be of sixteen inches thickness, the forms should be set up allowing for a space of that width.

The foundation, on the sub-foundation that has been previously wetted, should be composed of a mixture of one part Portland cement, three parts clean, sharp sand, and six parts of broken stone. The foundation ought to be four feet deep. Better satisfaction will be secured if provision is made for an air space between the walls. Two 6-inch walls 4 inches apart, and bound together with rods, will be a good arrangement. Separate forms for each must be constructed. Walls of this width will require no other reinforcement than the binding rods, provided the house is not to be high. One part of Portland cement, two of sand, and four of crushed stone will be the proportions of the mixture for the walls. The walls should be built in sections about two feet high at a time. Place the rods of half-inch iron with strong heads imbedded about two inches in the inner surface of each wall, and about a foot apart. This will help the wall stand the lateral pressure of any pile of ice within that may rest against it.

The roof, reinforced with ¾-inch iron rods a foot apart, is the next step. A form is constructed, of the desired angle. Upon this, about three inches of sand is placed and smoothed off carefully. Lay the rods so that they will rest one and one-half inches above the sand and put on a coat of three inches of concrete. The forms should not be touched for two weeks. Then the sand can be let out from the interior. All the openings between the walls and the roof must be closed up.

Small Storage Buildings. These should either be built of hollow concrete blocks, or, if monolithic in form, there should be an air space

in the walls. The air space is necessary to retard dampness and keep out the frost.

Root Cellar. Cellars of this character
usually are built half below and half above the
surface of the ground. When properly made,
they ought to be proof against all inroads by
frost. Excavation should be carried down to a
point sixteen inches below the desired level of
the floor. The foundation should be twelve
inches wide; and the mixture for it should be
one part Portland cement, three parts coarse
sand, and six parts gravel or crushed stone.
Forms should be made for the foundation.
After it is laid, the forms should be removed;
and a porous material, either broken bricks or
cinders, should be filled in over the floor space
to a depth of twelve inches. This should be well
tamped. The floor, of a total thickness of four
inches, should consist of three inches of concrete
and one inch of cement mortar. The concrete
should be one part cement, three parts sand, and
five parts gravel or crushed stone.

The wall of the cellar should be eight inches
thick, started from the center of the 12-inch
foundation. The mixture for the wall may be
one part Portland cement, two and one-half
parts sand, and five parts crushed stone, gravel
or cinders. Build up the end walls so as to form
a point at the middle, and high enough to give
the roof a sufficient pitch to shed the rain. Near
the top at each end, it should be remembered to
provide openings for windows; and the sash

should be fitted and plastered in after the concrete has set and forms have been removed.
Bins should be built of a size to suit convenience,

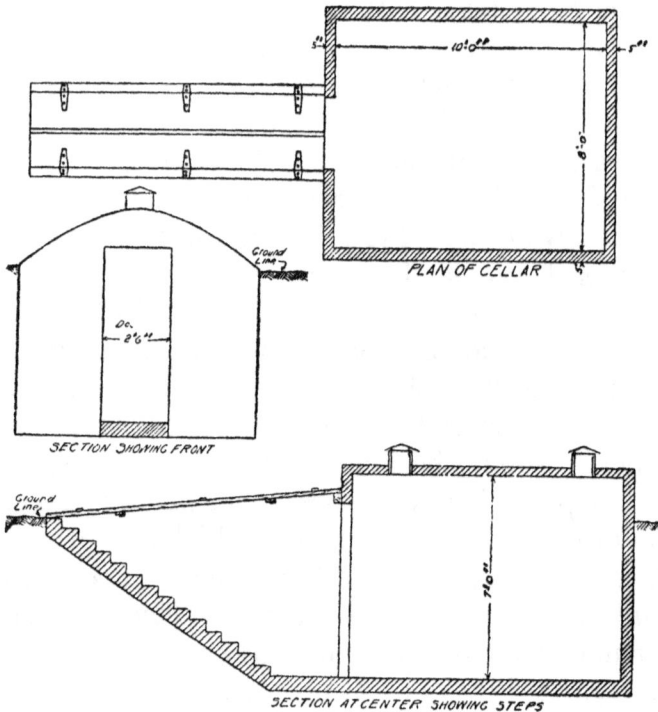

Fig. 31. A Cyclone Cellar.

with walls four inches thick and reinforced with
one-quarter-inch rods placed twelve inches apart
horizontally and vertically to give the bin walls
strength to withstand the lateral pressure when
they are filled with vegetables.

If a concrete roof is desired, forms should be
erected, and a roof two and one-half inches thick
built. On the top, and before the concrete is

dry, a quarter-inch layer of mortar, consisting of one part Portland cement and one part sand, should be placed and well troweled. The forms should not be removed before three weeks. The roof should be reinforced with woven wire fabric, and so should the steps. If the roof is sufficiently long to require supports, an eight-inch pillar in the center may be erected, reinforced with one-half-inch rods two inches apart and one inch from the surface.

Mushroom Cellar. The method of constructing a cellar of this character is the same as for a root cellar, with the exception that no floor will be required, and there should be little light.

Cyclone Cellar of Concrete. After fire, a farmer on the western plains fears a cyclone more than anything else on earth. The only sure salvation is to get below the surface of the ground, and practically every farm has a **cyclone cellar.** It is becoming the practice to construct these places of refuge of concrete. In using concrete as a building material, there is no danger of the roof blowing off, or of the walls rotting out in a few years and having to be renewed.

CONCRETE FENCE-POSTS

There is a constantly increasing demand for some kind of fence-post which is not subject to decay. The life of wooden posts is very limited, and the scarcity of suitable timber in many localities has made it imperative to find a substitute. A fence-post, to prove thoroughly satis-factory, must fulfil three conditions:

(1) It must be obtainable at a reasonable cost;

(2) It must possess sufficient strength to meet the demands of general farm use;

(3) It must not be subject to decay, and must be able to withstand successfully the effects of water, frost, and fire.

Although iron posts of various designs are frequently used for ornamental purposes, their adoption for general farm use is prohibited by their excessive cost. Then, too, iron posts ex-posed to the weather are subject to corrosion, to prevent which necessitates repainting from time to time; and this item will entail considerable expense in cases where a large number of posts are to be used.

At the present time the material which seems most ready to meet these requirements is rein-forced concrete. The idea of constructing fence-posts of concrete reinforced with iron or steel is by no means a new one; but on the contrary such posts have been experimented with for years, and a great number of patents have been issued covering many of the possible forms of rein-forcement. It is frequently stated that a rein-forced concrete post can be made and put into the ground for the same price as a wooden post. Of course this will depend in any locality upon the relative value of wood and the various materials which go to make up the concrete post; but in the great majority of cases, wood will prove the cheaper material in regard to first cost. On the other hand, a concrete post will last indefinitely, its strength increasing with age, whereas the wooden post must be replaced at short intervals, probably making it more expensive in the long run.

In regard to strength, it must be borne in mind that it is not possible to make concrete posts as strong as wooden posts of the same size; but since wooden posts, as a rule, are many times stronger than necessary, this difference in strength should not condemn the use of reinforced concrete for this purpose. Moreover, in many cases strength is of little importance, the fence being used only as a dividing line; and in such cases small concrete posts provide ample strength, and present a very uniform and neat appearance.

In any case, to enable concrete posts to withstand the loads they are called upon to carry, sufficient strength may be secured by means of reinforcement; and where great strength is required, this may be obtained by using a larger post with a greater proportion of metal and well braced as is usual in such cases.

In point of durability, concrete is unsurpassed by any material of construction. It offers perfect protection to the metal reinforcement, and it is not itself affected by exposure, so that a post constructed of concrete reinforced with steel will last indefinitely and require no attention in the way of repairs.

Reinforcement. No form of wooden reinforcement, either on the surface or within the post, can be recommended. If on the surface, the wood will decay; and if a wooden core is used, it will in all probability swell by the absorption of moisture, and crack the post. The use of galvanized wire is sometimes advocated; but if the post is properly constructed and a good concrete used, this precaution against rust will be unnecessary, since it has been fully demonstrated by repeated tests that concrete protects steel perfectly against rust. If plain, smooth wire, or rods are used for reinforcement, they should be bent over at the ends or looped to prevent slipping in the concrete. Twisted fence wire may be obtained at a reasonable cost and is very well suited for this purpose. Barbed wire has been proposed, and is sometimes used, although the barbs make is extremely difficult to handle.

For the sake of economy the smallest amount of metal consistent with the desired strength must be used; and this requirement makes it necessary to place the reinforcement near the surface, where its strength is utilized to the greatest advantage, with only enough concrete on the outside to form a protective covering. A reinforcing member in each corner of the post is probably the most efficient arrangement.

The Mixture. The concrete should be mixed with Portland cement in about the proportion $1:2\frac{1}{2}:5$, broken stone or gravel under one-half inch being used.

How to Preserve Gate-Posts. A sagging gate-post, rotted at the ground line, gives the whole place an unkept and rough look; and yet gate-posts do rot, and have a habit of getting out of line. The remedy is a very simple one, and very inexpensive as well. First brace the post

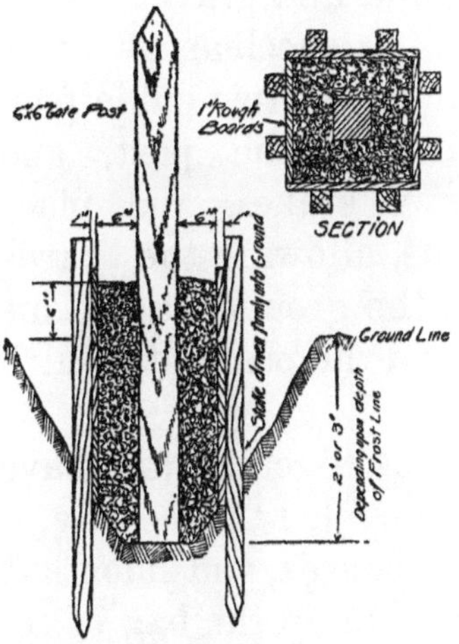

Fig. 32. Method of Preserving Gate-Posts.

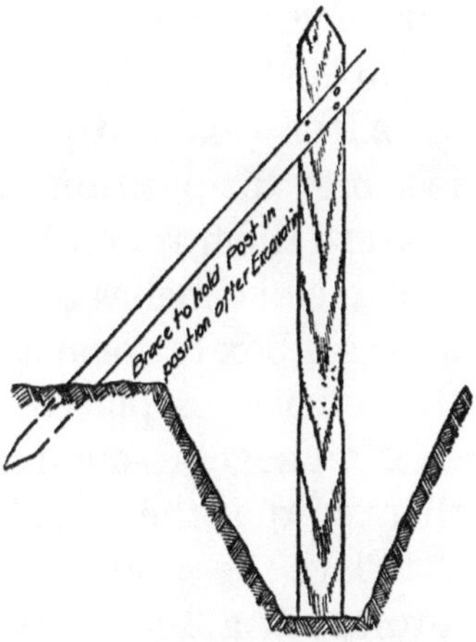

Fig. 33. Method of Preserving Gate-Posts.

in such a way as to prevent its falling. Then excavate around it, to a depth below the frost line. Pull the post into the proper position, and re-nail your braces. Fill the hole with concrete, to a point six inches above the ground, and your post will remain constantly in position. It will also last for years. When it has to be renewed, the old post can easily be pulled out, and a new one slipped into the hole in the concrete.

The materials needed for these repairs are:
1 Bag of Portland cement.
3 Cubic feet of gritty, clean sand.
6 Cubic feet of gravel, of a small size.

The cement can be obtained from a dealer in the nearest town. The sand and gravel can be obtained from your farm. A wheelbarrow holds about one and three-quarters to two cubic feet.

After excavating around the post, and bracing in position, drive stakes, and place against them rough boards, allowing the boards to come six inches above the ground line. This makes a box or form around the post, into which the concrete is placed.

Nail several small boards together, and have them so that they can be shoveled from easily.

Place the sand on these boards, and smooth it into about a 3-inch layer. Dump the bag of cement on the sand, and mix the same together thoroughly while dry. Smooth out this mixture, and shovel the gravel on top. This gravel should be thoroughly wet first. Then shovel the whole mixture from one pile to another, adding water enough to make a soft mass, turning over at least three times. Shovel the whole directly inside the box, and tamp with a 3-inch by 4-inch piece of lumber. When filled, take a trowel and smooth off the top, leaving the whole mass slightly higher against the sides of the post than at the edges.

After two days, remove the braces and the forms, and fill with earth around the concrete, up to the ground level.

It will take one man about half a day to do all the work. The materials, except the cement, will cost nothing. The cement will cost not over fifty cents a bag. Following these instructions will give altogether a very cheap and lasting job.

MOULDS FOR ORNAMENTS

In making artificial marble in moulds, we depart from the dry tamp process, unless iron

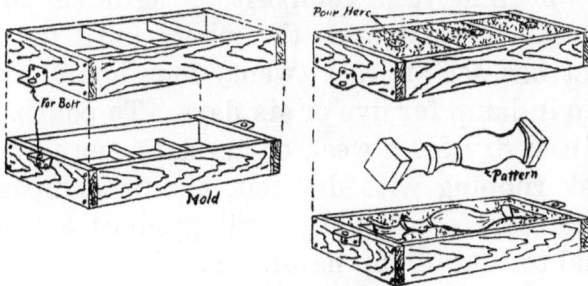

Fig. 34. Mould for Concrete Ornamental Shape.

moulds are used; but where only one or two pieces of each design are required, the cost of special iron moulds would bar their use.

However, should any iron moulds that meet the requirements be on the market, their use is preferable to that of wooden moulds. The latter may be made and used as follows:

Make a wood pattern exactly like the article desired, and a flask for it as shown, and, in fact, the same as is done in any cast-iron foundry. Fill the mould in the same way, only that one pound of plaster of Paris to every 12 pounds of sand is used. This composition is dampened and rammed into the mould. After it has been carefully filled and the wood pattern removed, let it stand for a day, then coat the interior with a heavy lather (lacquer) to insure a smooth finish. Close the mould, and fill with a composition consisting of one part Portland cement, one part marble dust, and two parts fine sharp sand, all to be mixed with sufficient water to admit pouring, keeping the composition agitated until poured. Do not open the mould until the cement has set at least twenty-four hours; then keep it damp for five or six days. To polish, let the cast dry for a week or ten days more, when brisk rubbing with dry felt, adding dampened oxalic acid occasionally, will produce a luster equal to that of any natural marble.

Openings for pouring must be made at all high places; and, before removing the pattern, it is well to penetrate the moulding sand with a steel wire, thus allowing a free escape of the air, thereby preventing honeycombs. See Fig. 34A.

Some experience is necessary to make this process a success, and by careful usage the same mould can be used several times. Crushed lime- will produce whiter but less durable work. stone may be substituted for the sand, which Almost any color can be produced by adding the coloring to the composition in the dry state. This process of making casts is adaptable to all manner of round spindles for stairs and porches.

Sidewalk Construction

It may be said that cement sidewalks are now taking the place of all other kinds of walks in all cities and towns of the country. They provide an even surface for pedestrians, are permanent, and present a pleasing appearance. The cost of the concrete sidewalk makes it competitive with all other materials; but in constructing such walks the maker must have an intelligent knowledge of the process in order to assure a walk that will be satisfactory. C. W. Boynton, chief inspector of one of the principal cement manufacturing companies in the United States, in a treatise on the subject, says:

"There are certain rules which should be observed in all cases, and in some cases additional precautions are necessary. The location of a walk is determined regardless of the natural fitness of the foundation, the soil and drainage conditions; and it is important, therefore, that these matters be carefully studied. The materials available should also receive careful attention, and should be selected with reference to quality, and not altogether with reference to cost. The weather condition at the time a walk is constructed has a marked effect upon its behavior, and must be taken into account to assure permanence.

"Poor workmanship, which includes im-

INTERIOR OF CONCRETE COW-BARNS, UNIVERSITY OF
TENNESSEE.
PLATE 10—CEMENT CONSTRUCTION.

proper proportioning of materials, the placing of a walk on an improperly prepared foundation, and failure to take into account weather conditions, are responsible for practically all failures that occur. Failures which can be positively charged to poor materials are few, though frequently materials are used which could be improved by a more careful and intelligent selection, without adding to the cost of the finished work. The quality of the work should always receive first consideration, and first-class materials should always be used, even though the cost is somewhat increased by the use of such materials. The principal expense in this class of construction is the labor item, and the labor will be the same whether good or poor materials are used.

"It is not uncommon to hear, as an argument to prove proper workmanship in a defective walk, that the same workmen had laid satisfactory walks by the same methods. Granting that the same workmen may have built two walks in the same manner, does not prove that the necessary precautions are always observed. The conditions might vary materially; and unless all points receive due consideration, success could result in one case, while failure would follow in another. The necessary qualifications, therefore, for the construction of sound and lasting work are **good materials, proper methods,** and **careful workmanship.** Failure to provide these will often result in disappointment."

CEMENT

In all sidewalk work, Portland cement is used. In small jobs, it is only necessary to secure cement from a reputable manufacturer; but where the quantity of work will justify, it is advisable to have the cement tested. The standard methods for testing, adopted by the American Society of Civil Engineers, and presented elsewhere in this work in the section on the "Materials and Manufacture of Concrete," should be used; and the cement should comply with the Standard Specifications for Cement of the American Society for Testing Materials, which will also be found in the same connection.

AGGREGATES

The first requisite in the selection of the stone for the concrete is **cleanliness.** This is absolutely essential to strength in the concrete. In selecting an aggregate, the character of the surfaces presented by the particles should always receive close attention; these must be hard and permanent. A covering of any fine material will interfere with the cement or mortar getting into contact with the surface of the aggregate, and the strength will be reduced proportionately. An excellent precaution in this respect is to **avoid the use of dirty materials.**

Some experimenters found that certain sands gave better strength with the addition of 10 or 15 per cent of finely divided clay than when tested without the clay. This, however, is no argument in favor of dirty materials. The addition of a small percentage of finely divided clay might be permissible when the clay is treated as a separate material, while even a much smaller quantity naturally occurring in the aggregate might make it wholly unfit for concrete purposes.

In order to obtain the best results, the **aggregates should be well graded;** that is, they must not contain an excess of one-size particles, and must contain but a small percentage of fine particles. In the case of stone, the material will usually be quite satisfactory, provided the stone in itself is hard and durable and not affected by exposure to the elements, and provided it is prepared and marketed under conditions which assure its being clean and free from a covering of dust or other matter.

Some stone, though apparently quite hard, presents a chalky surface with which it is impossible for the cement to form a perfect bond. Stone of this character should be avoided, for it cannot possibly produce good concrete.

In sand and gravel, one is dealing with entirely different materials, but materials probably to be preferred to stone and screenings, when selected with sufficient care. The use of sand and gravel is very popular, owing to the ease with which they are obtained in many localities. Where these materials are readily secured, they are frequently used as they come

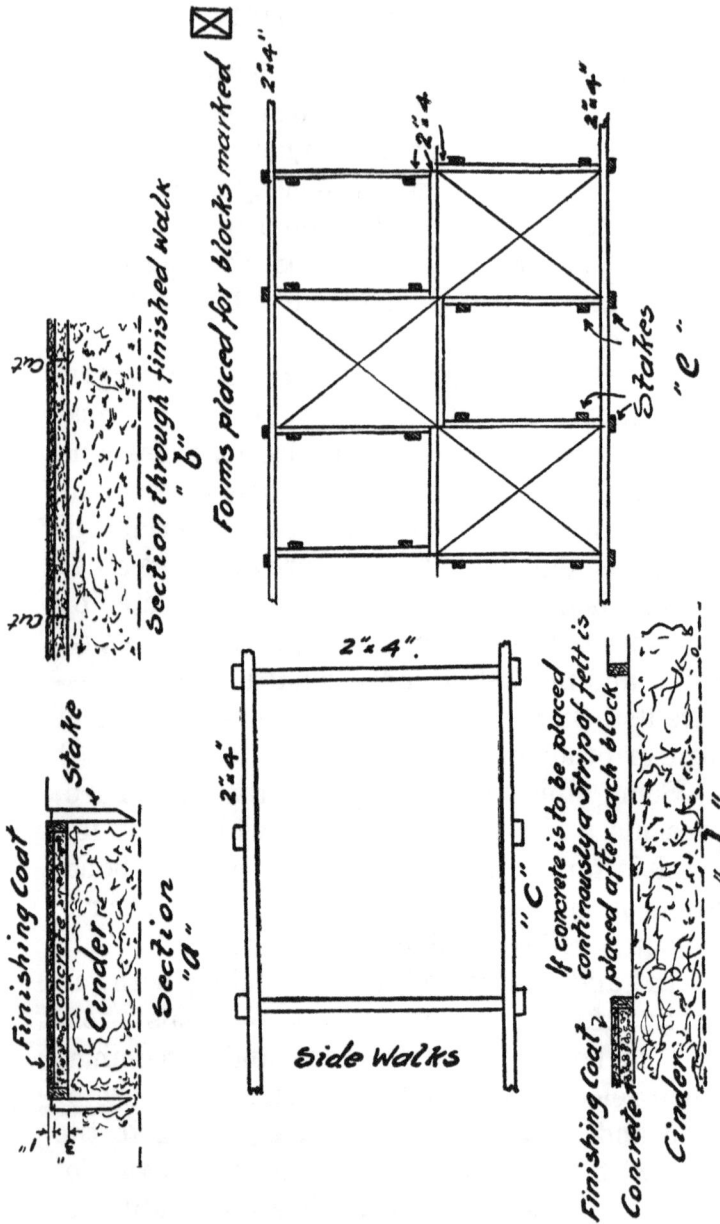

Finishing Coat

Stake

Cinder

Section "A"

2"x0"

2"0"

Section through finished walk "b"

Forms placed for blocks marked ⊠

2"x4"

2"x4"

2"x4"

Stakes

"C"

2"x4"

2"x4"

"C"

Side walks

If concrete is to be placed continuously a strip of felt is placed after each block

Finishing Coat

Concrete

Cinder

"b"

Fig. 35.   Detail of Concrete Sidewalk Construction.

from the deposit, with little or no thought given to their fitness for the work in hand. The character of the materials which are sometimes used in concrete is surprising. Aggregates should always be firm and hard, and should remain so when exposed indefinitely to the weather. It is quite common to find a considerable quantity of shaly pebbles in some of the glacial sands and gravels of the upper Mississippi Valley. These pebbles are not strong in the first place; and they disintegrate readily when exposed to the elements. They also absorb water readily when used in concrete, and expand under the combined action of moisture and frost, injuring the concrete to a greater or less extent. Though the effect of the soft sand grains is not so apparent as is the effect of the larger pebbles, such sand cannot possibly produce first-class results if the shaly particles form any considerable portion of the sand content. In the territory referred to, it is not unusual to find sidewalks badly pitted and marred, as a result of the disintegration of this shaly material. These shaly particles are undesirable, because they are both weak and unstable. **A concrete can never be stronger than the material making up the aggregate.**

The size of the sand grains and the relative proportion of grains of different size, have a very marked effect on the value of the sand. At least 75 per cent of a sand should be retained on a 40-mesh sieve, with the particles well distributed between that size and the size passing a 4-mesh sieve, with an increasing proportion on the coarser sieves. Such a sand will have much less total surface than one composed of equal proportions of particles on the several sieves. A sand made up entirely of fine particles will present a very much larger surface which must be covered with cement, than either of the sands above mentioned. For instance, the total superficial surface of a given volume of spheres one-sixty-fourth inch in diameter is sixteen times the surface of the same volume of spheres one-fourth inch in diameter. As the making of a first-class concrete necessitates the perfect covering of every particle of sand with cement, and every particle of the coarser aggregate with the cement-sand mortar, it is apparent that materials with an excess of fine particles should be avoided. The same line of reasoning is applicable to the combined aggregate in the concrete.

Occasionally one sees a mixture of cement and sand used for the concrete base in sidewalk construction, and cannot help being impressed with the fact that the user fails to appreciate tne requisite of a good concrete.

In Table VI will be found a physical analysis of a material taken from a sidewalk job, in which it was being used for the concrete in the proportion of one part cement to four parts aggregate.

The general quality of this material was fairly good, though it will be noted that only 2 per cent of it could be considered gravel. No particles found in the sample were larger than

TABLE VI

Sand Analysis

Weight of Sample—500 Grams										Measured and Computed Voids		Specific Gravity
Percentage Retained on Sieve No.									Through No.	Measured	Computed	
4	10	20	30	40	50	80	100	200	200			
2.0	15.6	14.8	17.8	11.2	5.4	27.6	2.2	3.8	1.0%	29.2	33.2	2.614

½ inch. The computed voids in the sand were 33 per cent; measured by displacement, 29 per cent. The mixture of 1 cement to 4 sand, therefore, was out of balance, the cement not being sufficient to fill the voids. Not only did this volume of sand contain more voids than the cement could fill, but the excess of fine material detracted from the value of the sand as it was being used, because it presented a very much larger surface than the cement could possibly cover.

Size of Aggregate. Aggregates exceeding 1¼ inches in diameter should not be used. Undoubtedly there are many gravels which would give good results, though containing larger sizes; but this limit is safe and the one most often applied to this class of work. The lower limit, ¼ inch, which is also the upper limit for sand and stone screenings, is almost universally accepted.

Unscreened Gravel. In many districts, unscreened gravel (gravel as it comes from the bank containing both coarse and fine particles) is used. This practice should be avoided, as such material usually contains a large excess of sand, and would be much improved if screened and the proper proportions of fine and coarse particles remixed. The increased value of the remixed aggregate over the natural material would more than justify the additional expense. The case referred to in Table VI illustrates this point quite clearly.

FOUNDATION OR SUB-BASE

The foundation must provide a permanent bed for the walk, and serve as a means for disposing of water which would otherwise accumulate under the walk. In many localities, a well-constructed sub-base will offer sufficient drainage; but in some soils and under some conditions additional drainage is necessary.

Drainage. If water is allowed to accumulate in the sub-base, there is danger of the walk being heaved by frost. Therefore, in soil where the sub-base and the natural drainage cannot take care of the water, other drainage should be provided. The best means of supplying this additional drainage will depend somewhat upon the available outlets, etc. In some cases stone-filled trenches, properly placed at intervals along the walk, will provide adequate drainage, while in other cases a tile drain will be necessary.

Material. The material to be used for the foundation or sub-base of a walk will depend to a great extent upon the locality in which the

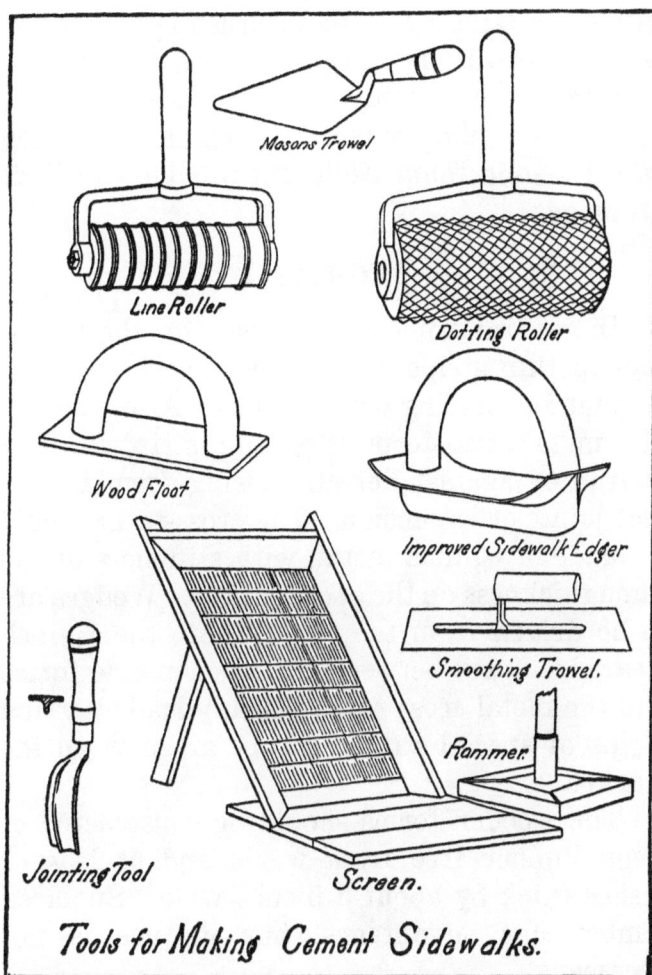

Masons Trowel

Line Roller

Dotting Roller

Wood Float

Improved Sidewalk Edger

Smoothing Trowel.

Rammer.

Jointing Tool

Screen.

Tools for Making Cement Sidewalks.

Fig. 36. Tools Especially Used in Making Cement Sidewalks.

work is contemplated. The builder can best determine from the materials available which one is the most satisfactory and economical. The one chosen must be of such a character as to withstand tamping, without crushing to the extent that it will prevent proper drainage. **Steam cinders** are commonly used for the sub-base; and if the fine material is eliminated, they afford a solid foundation and provide excellent drainage.

FORMS

In general, wood will be used for the forms, though thin strips of metal will be found convenient in forming curved lines. Also, the use of a metal cross-form or parting strip will be a guarantee against defects arising from imperfect joints or expansion. The cross-form should be made of 1/8-inch metal, with stiffeners of the same thickness on the ends and top. Wedges are to be driven from the outside into the 1/2-inch clearance space between the wooden side forms and the metal cross-form. Ready-made parting strips of special patented type are now on the market.

The wooden forms should be constructed of clean lumber free from warp, and at least 2 inches thick by about 5 inches wide. Surfaced lumber has advantages, but its use is not necessary.

In placing the side forms along the line of the walk, care should be taken to maintain a good alignment, and they should be leveled so as to conform with the finished grade.

Providing for Surface Drainage of Walk. The form nearest the street should be slightly below the inside form, thus providing a drain which will prevent water from collecting on the walk. The side forms should be securely staked, the stakes alternating on either side about every two feet. If the special metal cross-form is used, fewer stakes will answer, for when the form is keyed into position, it is rigidly fastened and holds the outside forms in their proper relative position. Wooden cross-forms need only be held in place by stakes on the opposite side from which the concrete is to be deposited. When the concrete is being placed, a shovelful or two will hold the cross-forms firmly until it is tamped into position.

When wooden cross-forms are used, the location of the joints should be definitely determined and plainly marked on the side forms before any concrete is placed. The cross-forms should be placed so that the face against which the concrete is to be packed is in line with the points indicating the position of the joints.

Providing for Expansion Joints. About every 50 feet one of the wooden cross-forms should be replaced by a metal parting strip, which should be left in the walk until it is opened to traffic, when it will be removed and the opening thus produced filled with paver's pitch or other suitable material. This forms an expansion joint, which insures the walk against cracking. This precaution is also necessary when a new walk abuts curbing or other cement or stone walk.

SURFACE TREATMENT

The surface treatment which a walk receives depends largely upon the practice in the community in which the work is being done. The smooth, **steel-trowel finish** is probably the most common and at the same time the poorest finish used. Such a finish frequently results in crazing or hair-checking of the surface, which is due to nothing more than a slight contraction which takes place in the film formed on the surface by the steel trowel. Besides the smooth finish showing every little blemish and variation in color, it is much more slippery than any of the other finishes.

The **wooden-trowel finish** is growing in popularity, and certainly has many points in its favor. The **brush finish** is similar to the wooden-trowel finish, but it requires an additional tool, and one that can be used for no other purpose. The finishes that are produced by special tools, like the tooth-roller, etc., have little to commend them. They are in no way superior to the rough finish produced in a simpler manner, and do not harmonize so well with the usual surroundings.

Marking. There might possibly be some chance for argument regarding surface finish, but certainly surface marking will not permit of any. The position of the joints between the blocks should be determined before the base is placed, and provided for in the construction.

Positive joints should always be provided in the base of the walk. These are the real joints, and the markings in the top should always occur over them. It is not sufficient to make a surface marking, together with a feeble effort toward cutting through the base with a small trowel or similar instrument. **More walks are disfigured by failure on the part of the builder to provide proper joints than by any other cause.**

Size of Block. The size and shape of the blocks into which a walk is divided are governed very largely by the width of the walk, the local practice, and personal tastes. Other points, however, should be considered; in fact, local practice and personal tastes should be eliminated entirely when walks on business streets are being constructed. Where the whole space between the building line and the curb is to be covered, many angles and irregular lines are introduced, owing to openings, steps, etc. Steps should never be constructed over a joint; nor should a joint ever be permitted to intersect a step (excepting at a joint), unless the walk and step are constructed entirely independent of each other. Joints between the blocks should be placed so as to avoid small corners and unnecessary angles; in fact, so far as possible, all blocks should be rectangular. Also the joints in new work, abutting old, should always be projected from the joints in the original work, unless a distinct open joint is provided between the new and the old.

YIELD OF ONE BARREL OF PORTLAND CEMENT

A barrel of Portland cement should contain 380 pounds net, or three and one-half cubic feet. One operator states that if work is well tamped, a barrel will yield the following:

	CEMENT.	SAND.	GRAVEL.	THICKNESS.	PER BBL.
Concrete course	1	1	6	3 inch	
Top course	3	5	.	1 inch	52 sq. ft.
				4 inch walk	
Concrete course	1	1	6	3¼ inch	
Top course	3	5	.	¾ inch	55 sq. ft.
				4 inch walk	
Concrete course	1	1	6	3½ inch	
Top course	3	5	.	1 inch	49 sq. ft.
				4½ inch walk	
Concrete course	1	1	6	4 inch	
Top course	3	5	.	1 inch	42 sq. ft.
				5 inch walk	
Concrete course	1	1	6	4½ inch	
Top course	3	5	.	1½ inch	31 sq. ft.
				6 inch walk	
Concrete course	1	1	6	6½ inch	
Top course	3	5	.	1½ inch	24 sq. ft.
				8 inch walk	

CAUSES OF DEFECTS

Frozen concrete surfaces have the appearance of being spotted. A surface frozen before the concrete has set, scales off easily. Paper, tarpaulin, or clean sand can be used to prevent freezing. Good cement work can be done in freezing weather by using warm water, warm sand, and gravel, and protecting the material from freezing.

Sunburned surfaces have the appearance of over-burned clay. Good results can be secured in hot weather by covering with canvas.

Poorly mixed cement surfaces look streaky when set. Dirty streaks are caused also by the use of **unclean materials.**

By using **too much water** in the top mix, the cement is washed out, and a good **union** between the top and concrete is not obtained. The surface coat is also prevented from adhering by dirt or by weak concrete.

Insufficient tamping makes the work porous and weak. The top surface should be laid before the under concrete is set.

Over-troweling the surface coat causes hair-cracks and work that is slippery, rough, and wavy.

Cracks are caused by poor mixing, by too much tamping, by shocks in the early stages of setting, by poor concrete, by walking on scantling when work is new, by making the blocks too large for the thickness of the work, by roots of trees growing under the walk, or by not cutting work through at joints to allow for expansion.

Pock marks are caused by **improper floating before marking. Rain** on the work before the top has set, also causes pock marks.

Sloughing off is caused by insufficient cement, poor mixing, or the use of poor material.

Too much water causes honeycombed and streaky work, and also is apt to prevent a good bond between the top and base. Insufficient water will affect the strength of the concrete.

CURBS AND GUTTERS

The foundation for curbs and gutters, as for sidewalks, should be governed by the soil and climate.

Concrete curbing should be built in advance of the walk, in sectional pieces 6 feet to 8 feet long, and separated from each other and from the walk by tar paper or a cut joint, in the same manner as the walk is divided into blocks.

Curbs should be 4 inches to 7 inches wide at the top, and 5 inches to 8 inches at the bottom, with a face 6 inches to 7 inches above the gutter. The curb should stand on a concrete base 5 inches to 8 inches thick, which in turn should have a sub-base of porous material at least 12 inches thick. The gutter should be 16 inches to 20 inches broad, and 6 inches to 9 inches thick, and should also have a porous foundation at least 12 inches thick.

Keeping the above dimensions in mind, excavate a trench the combined width of the gutter and curb, and put in the sub-base of porous material. On top of this, place forms, and fill with a layer of concrete, one part Portland cement, three parts clean, coarse sand, and six parts broken stone, thick enough to fill the forms to about 3 inches below the street level. As soon as the concrete is sufficiently set to withstand pressure, place forms for the curb; and, after carefully cleaning the concrete between the forms and thoroughly wetting, fill with concrete,

INTERIOR OF COW-BARNS, UNIVERSITY OF TENNESSEE.
PLATE 11—CEMENT CONSTRUCTION.

Johnson or Corrugated Bar.

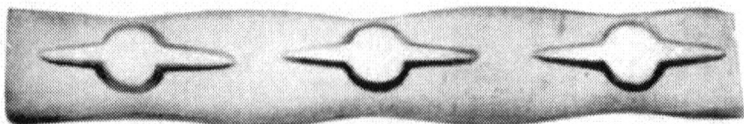

Thacher Bulb Bar.

Diamond Bar.

Ransome Twisted Bar.

Cup Bar.

Universal Bar.

Twisted Lug Bar.

TYPES OF DEFORMED BARS.

PLATE 12—CEMENT CONSTRUCTION.

one part Portland cement, two and one-half parts clean, coarse sand, and five parts broken stone. When the curb has sufficiently set to withstand its own weight without bulging, remove the ¾-inch board between the face of the curb and the form (shown in Fig. 37), and with the aid of a trowel fill in the space between the

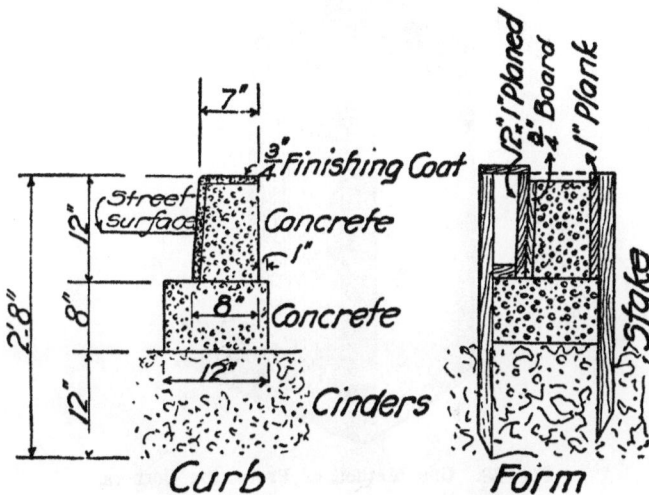

Fig. 37. Concrete Curb Construction.

concrete and the form with cement mortar, one part Portland cement and one part clean, coarse sand. The finishing coat at the top of the curb should be put on at the same time. Trowel thoroughly and smooth with a wooden float, removing face form the following day. Sprinkle often and protect from sun.

In making curbs alone, excavate 32 inches below the level of the curb, and fill with cinders, broken stone, gravel, or broken brick, to a depth

of 12 inches.   Build a foundation 8 inches deep
by 12 inches broad, one part Portland cement,
three parts clean, coarse sand, and six parts
broken stone; and from the top of this and
nearly flush with the rear, build a concrete wall
$11\frac{1}{4}$ inches high, $7\frac{1}{4}$ inches broad at the base,

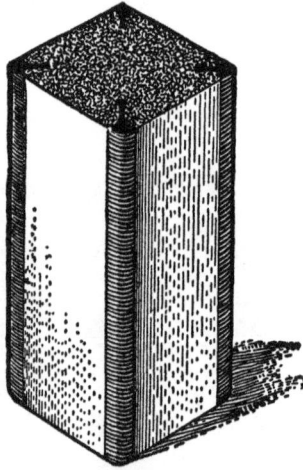

Fig. 38.   One Method of Protecting Corners.

and $6\frac{1}{4}$ inches at the top, the 1-inch slope to be
on the face.

Remove the forms as soon as the concrete
will withstand its own weight without bulging,
and put on the finishing coat in the manner as
indicated above.   Keep moist for several days,
and protect from the sun.   Measurements may
be varied to suit local conditions.

Protecting Corners.   It is often necessary to
protect the exposed corners of concrete work.
For this work there are a number of devices.

One is the "Wainwright" system, in which the protective material is a galvanized steel T-bar with a dovetailed web.

CONCRETE TILE

Cost of Concrete Tile. The cost of producing tile is not the same in all sections, owing chiefly to variations in freight rates. One estimate, based on a machine cost, is as follows:

SIZE OF TILE.	COST PER THOUSAND.	PRICE PER THOUSAND.
4-inch	$13.00	$22.00
5-inch	15.00	28.00
6-inch	19.00	38.00
7-inch	27.00	48.00
8-inch	36.00	60.00
10-inch	49.00	90.00
12-inch	67.00	115.00

In the making of concrete tile, the same general principles apply that rule in other concrete work. It is absolutely necessary to use only the best of materials.

Sand. This should be clean and sharp, and should be well graded. Large stones are not advisable. No loam should be used, but a very small quantity of clay is allowable, and helps to make the product waterproof.

Cement. The best cement, that which will pass the test of the American Society for Testing Materials, is recommended.

Mixing. Whether the mixing is done by machinery or by hand is of little consequence, so long as the operation is thorough. If tile in any large quantity are to be made, the machine method will be found more economical. The vital point about concrete is to produce a tile that will hold water, and the whole matter depends more upon the mixing than upon anything else, although there should be no skimping of cement.

Machines. There are several good tile machines on the market that have proved their worth in actual practice.

In making the smaller sizes of tile, some machines use no pallets, the moulds holding the tile until they can be placed upon boards, on cars or racks, when the mould is released. Where it is possible, use cars to carry the product from the machine. It is probable that the car system is preferable to the rack system for handling the product, as it saves handling the tile twice, and the convenience of cars is increased where steam-curing is practiced. The number of tile handled naturally depends entirely upon the capacity of the machine used and the number of men employed. The average machine, with a crew of seven men, will turn out from 300 to 500 perfect tile an hour, in sizes ranging from 4 to 12 inches. It will be seen that with from 3,000 to 5,000 tile to dispose of in every working day of ten hours, the matter of racking them is of greatest importance, as time lost is money lost.

In using cars, the tile can be stacked three tiers high and run away from the machine to the curing room, to stand there until they can be run to the yard. In this way the tile are handled but twice—once to the cars from the machine, and once from the cars to the racks or ground pile in the yard.

If the tile are cured on racks, these can, of course, be built as high as it is considered practicable to reach.

A **steam-curing plant** may be arranged with little expense. Select a room that can be closed up tight, in the absence of a regular brick-walled kiln, and run your steam pipes under the floor. Use damp steam, turning it into the room under not more than 5 pounds pressure. The combination of warmth and moisture hastens the setting, and provides for perfect crystallization. Tile are of course much more quickly cured by this method than by any other. Steam is not turned onto the product until it is 12 hours old, and the tile are left in the kiln from 18 to 24 hours after the steam is applied.

A steam-curing kiln may be operated without a boiler, if it is necessary. Place thin sheets of sheet iron over gas jets or gasoline torches and let water drip onto them from above. The water is turned into steam as soon as it touches the hot iron and the resultant heat and moisture do the curing.

Ageing. Tile should be aged for from 30 to 60 days before being used. It is poor policy to attempt to rush matters and use tile before they are properly aged, as the tile are very apt to fail later because of this oversight.

Freezing.   It is impossible to harm a cement tile by freezing and thawing, after it is two days old.   Tile can be hauled with perfect safety in freezing weather, and they are not injured by being left on the ground throughout the winter, while clay tile would disintegrate under such conditions.

CONCRETE WALL BUILT TO SAVE A GRANITE WALL AT LOUISVILLE, KY.

The piles supporting the original wall rotted through alternate wetting and drying from freshets in the creek near base, causing the wall to start sliding. This was effectively stopped by the erection of a reinforced concrete wall of inverted T shape on reinforced concrete piles, about 10 ft. in front of the old wall as shown, and filling in behind with earth covered with dry block stone to prevent washing.

Cement Construction

Reinforced Concrete

Introduction. Reinforced concrete is concrete in which steel has been embedded to give additional strength and elasticity.

Steel has about the same strength in tension when used as a beam, as it has in compression when used as a column or post. The same thing is approximately true of wood and some other materials of construction. In concrete, however, the conditions are quite different, the compressive resistance of concrete being about ten times its tensile resistance.

In a concrete beam, the upper portion of the beam is in **compression**, and the lower part is in **tension**. The line where the internal stresses of the beam section change from compression to tension is called the **neutral axis**.

The forces must balance on each side of the neutral axis. A plain concrete beam, being so much stronger in compression than in tension, will have its neutral axis located very low.

Steel is so much stronger in tension than concrete that a very small steel rod or bar placed in the bottom of a concrete beam will raise the neutral axis and balance the compressive forces exerted above.

Plain concrete, when used in the form of pillars and posts, is capable of carrying heavy direct loads through its great compressive strength. But when it is subjected to a direct pull—that is, to tensile strains—it is weak. For example, if a plain concrete beam is subjected to a load, it will break apart at the bottom just as a piece of chalk would under like conditions, being unable to resist the tension in the lower portion of the beam. In order to overcome this, reinforcing steel is used to give proper tensile strength and elasticity. The concrete in the top of the beam takes care of the compression. A properly reinforced concrete beam, therefore, has the strength of stone in resisting compression, united with the tension-resisting power of steel.

When a beam is loaded and supported at the two ends, it will have a tendency to deflect or bend. To illustrate, assume that a beam is made up of a series of flat plates—or, in other words, like a pad of paper or a book—the difference being that in the pad of paper the leaves are not in any way connected to one another, whereas in a beam the adhesion or sticking together of the various particles of the material ties the imaginary plates together. Now, when the supposed beam starts to deflect, one of two things will happen: either the various plates separate, as when a book or pad of paper is bent, and, in separating, slide by one another; or, if the plates are held together and sliding is prevented, the particles in the upper plates compress, and those in the lower plates elongate or stretch out.

It is thus seen that in addition to the compression and tensile stresses in the top and bottom of the beam, there are internal stresses of

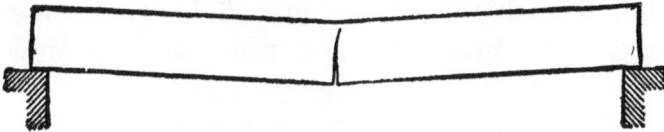

Fig. 39.   Plain Concrete Beam.

equal importance, against which the concrete must also be properly reinforced. These may be **tensile** or they may be **shearing** forces.

Fig. 39 shows a plain concrete beam, supported freely at the ends, which has failed upon

Fig. 40.   Beam with Horizontal Reinforcing Rods.

the application of a small load applied near its center.

Fig. 40 shows a similar beam having horizontal reinforcing rods located near the bottom surface of the beam. The method of failure under a medium load, in this case, was said to be due to the ends of the reinforcing rods **slipping** in the concrete. The diagonal or slanting cracks are partially due to horizontal shear set up by the bending of the beam. These are sometimes spoken of as due to **diagonal tension.**

A means of fortifying against horizontal shear is by the use of **stirrups**. Bands or rods of steel or iron are bent in the shape of a **U**, and either placed loosely around, fastened rigidly to, or made as a part of the reinforcing rods. The detail of this construction will be given later under the head of "Materials for Reinforcement" and "Reinforcing Systems."

Fig. 41 shows the method of failure of the same type of beam as previously shown, but having loose stirrups surrounding the horizontal

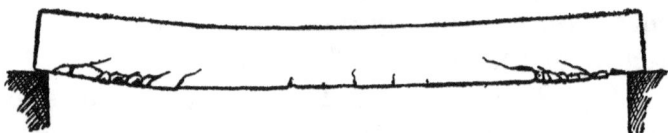

Fig. 41.   Beam with Horizontal Reinforcing Rods and Loose
Stirrups.

reinforcement bars and embedded vertically in the concrete. This beam failed when tested to destruction, by the slipping of the horizontal rods. The figure shows the shearing of the concrete **along** the horizontal plane above the rods, but no diagonal cracks. The stirrups evidently prevented the shearing action above the rods. As a means of preventing such a method of failure, some companies have either rigidly fixed the stirrups to their reinforcing bars or formed them as a part of same. The result of such a construction seems to throw the greater part of the body stresses of the beam onto the horizontal bars for support. Some authorities consider this a weakness in the construction.

Fig. 42 shows the failure of the same type of beam as previously shown, but provided with reinforcing bars with fixed stirrups. From the

Fig. 42. Beam with Horizontal Reinforcing Rods and Fixed Stirrups.

cracks shown near the bottom of beam, this seems to be a well-balanced reinforcement, with the main stress occurring at the center of the horizontal rods.

While these different forms of reinforcement seem to be favored by many, they are criticised by some investigators who claim to have records of tests showing that the additional strength of the stirrup construction does not make up for the additional cost. The method of reinforcing shown in Fig. 43 has been successfully used in

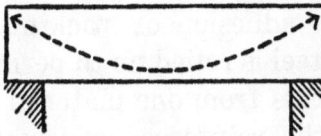

Fig. 43. Anchored End Bar Construction.

deep beams, and also in beams which are "continuous"—that is, which extend over more than one span between columns.

In Fig. 43, the method of fastening the ends of the reinforcement is a point to be noticed. In case of a continuous beam, the reinforcement would simply extend upwards toward the end of the column, and over into the next span. It is claimed by many authors that the ends of all rods for reinforcement should be either bent over and embedded in the concrete, or fitted with some kind of expanded end, to prevent the ends slipping when the beam is bent.

The most important principle in placing reinforcement in concrete beams is to place the steel so that it will relieve the concrete from all tensile stresses if possible, and thus aid in developing the high compressive strength of the material.

Every ounce of tension in the steel is only effective as it is transferred to the concrete. In the case of a plain beam with free ends, there is no stress in the steel at the ends, while the maximum tension is usually at or near the center of the beam. The entire amount of this tension must be gradually transferred from the steel to the concrete.

While the adhesion or sticking of the concrete to the steel is relied on to permit the transfer of this stress from one material to the other in much of the reinforced concrete work now being done, it is realized that this adhesion is not always permanent. Failures of floors have already occurred, due to loss of the adhesion, after they have successfully supported heavy loads for many years; the adhesion being greatly reduced with age and under certain unfavorable conditions, such as continued soaking of the concrete in water, long-continued vibration, etc.

Experience has demonstrated that beams may fail in other ways than by the pulling in two of the reinforcing steel, as, for example, by shearing across a vertical plane, by tension along a diagonal plane, or by slipping of the rods through the concrete.

HISTORY OF REINFORCED CONCRETE

The history of reinforced concrete seems to be a very uncertain quantity. Authors vary as to the date at which it made its first appearance; also, as to the form in which it first came into public notice. In fact, very little authentic history is available concerning its truly early stages. Its history since the nineties, however, is well recorded and is being recorded in enduring form by the increasing number of uses to which it is being subjected each year.

Some of the earlier records furnish interesting forecasts which are to-day becoming real and important factors in construction work. Louden, in his "Encyclopedia of Cottage and Villa Architecture," wrote:

"Floors and roofs might be made flat by means of a lattice-work of iron tie-rods, thickly embedded in cement or concrete and cased with flat tiles."

Here we see another forecast of reinforced concrete as practiced to-day.

A Frenchman, M. Lambot, patented in France in 1855 a system of reinforcement which he called a substitute for wood, and which consisted of a network or parallel set of wires, bars, or rods, embedded in or cemented together with hydraulic or other cementing matter, so as to form beams or planks of any desired size.

This was the same year that M. François Coignet, also a Frenchman, patented a system for making concrete, or beton, from hydraulic lime. In 1861 M. Coignet published a pamphlet advocating metal reinforcement, and described various ways of applying it for strengthening concrete floors. But although his system appears to have received some attention in France, it was not until 1879 that any work was carried on by him there.

The much earlier systems of Lecomte, Thusane, and others, of using rods and bars embedded in concrete, probably had something to do at least with suggesting these inventions. About 1860, Mr. Brannon, an architect and engineer, anticipated the use of wirework and iron in other forms as reinforcement, for he describes his invention as:

"A method of forming roofs, floors, ceilings, doors, walls, and other parts of buildings, or other structures, of cement or concreted materials in combination with metallic, fibrous, or laminated substances, with a view to render them more durable, fireproof, and healthy; and it consists in employing for the said purposes a sustaining metallic framework or skeleton, firmly fixed and bolted or bound together, upon which is stretched wirework, so as to partially enclose or be completely embedded in the said concreted materials which compose the body of a structure, or any part thereof, thereby perfectly bonding the same into a solid and coherent mass."

The system was, however, too costly and intricate to come into general use.

When, in the late sixties, M. Monier, a French gardener, began making flower-pots, boxes, and small water-tanks out of concrete, and embedded wire in the material to increase its strength and decrease its weight and bulk, he little thought that forty years later the principle which he used and upon which he was granted a patent, would be used throughout the entire world in the erection of millions upon millions of dollars worth of construction work.

One of the first uses of reinforced concrete in building construction was in the house erected by W. E. Ward in 1872, at Port Chester, N. Y. As previously mentioned, however, some twenty years earlier than this, in France, the first combinations of iron imbedded in concrete were made in a small way. But it was not until the very end of the last century, since 1895, that reinforced concrete came to be employed commercially in the construction of buildings. Previously to this the plain type of construction had attained a wide use in foundations, and at this time its development was beginning for such structures as dams, sewers, and subways.

Two principal reasons may be offered for this comparatively slow growth followed by such marvelous activity. In the first place, Portland cement manufacturers, beginning in Europe about the middle of the 19th century and in the United States about 1880, finally produced a grade of cement which, with the inspection necessary for all structural materials, could be depended upon to give uniform and thoroughly reliable results; furthermore, along with the perfection of the process of manufacture, the price gradually fell from the high cost per barrel in 1880 for imported cement, to a figure for domestic Portland cement of equally good, if not better, quality, at which concrete in plain form could compete with rough stone masonry, and, with steel embedded, could compete with other building materials.

In the second place, theoretical studies and practical experiments have now produced rational and positive methods for computing the strength of concrete reinforced with steel, so that absolute dependence can be placed upon it.

Briefly, reinforced concrete such as is used for construction of industrial buildings, bridges, retaining walls, etc., consists of Portland cement, sand, and gravel or broken stone, mixed with water to a consistency that will just flow sluggishly, and in which steel rods are embedded so as to produce an artificial stone having many of the characteristics of steel.

In the earlier stages of reinforced concrete, and even up to the present time, many patents of a more or less fundamental character have been granted. These have taken the line of special forms of reinforcing metal as well as methods of design. Some of the principal styles of reinforcement are illustrated below, under "Re-

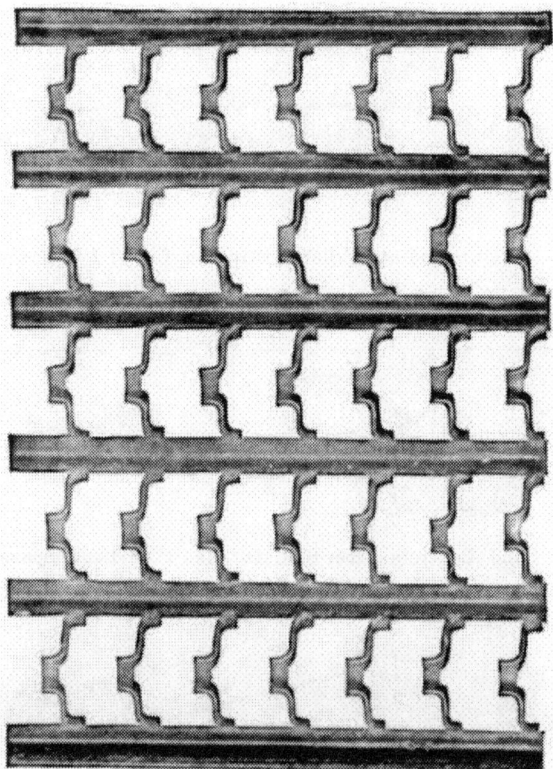

Rib Lath.

Truss Lath.
TWO FORMS OF METAL LATH.

PLATE 13—CEMENT CONSTRUCTION.

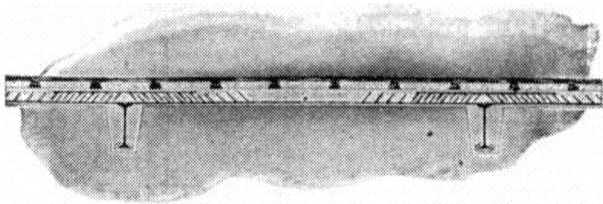

Reinforced Solid Slab Resting on Top of I-Beams.

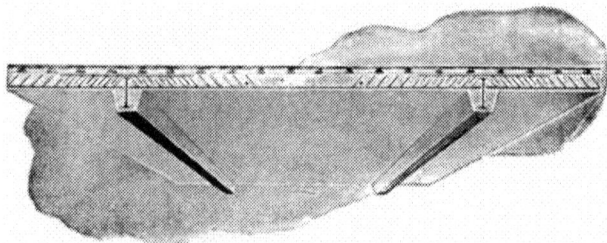

Solid Concrete Floor-Slab for Short or Long Spans.

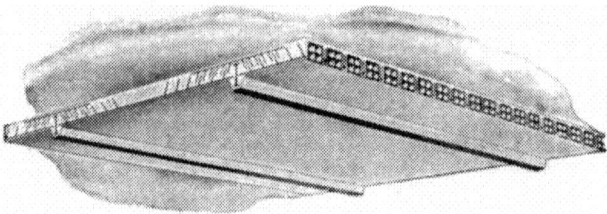

Reinforced Hollow Tile Long-Span Construction.

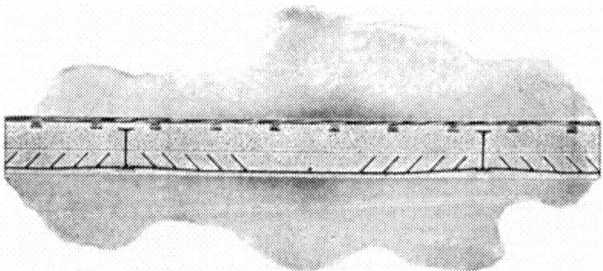

Slab Resting on Lower Flange of I-Beam. Note Flat Ceiling and Cinder
Fill over Concrete Construction.

DETAILS OF FLOOR CONSTRUCTION—FRAMING BETWEEN STEEL GIRDERS.

PLATE 14—CEMENT CONSTRUCTION.

inforcing Systems.'' While it is not necessary to encroach on any of these inventions in building, the field is worth careful consideration from the viewpoint of economy and durability, as to whether or not it may be advisable to make use of them.

There has been no class of structures, no line of the building trades, which has not been affected by reinforced concrete, and many of them have been revolutionized.

Relation to Portland Cement Industry. It is interesting to note the relation of the increased use of reinforced concrete construction to the growth of the Portland cement industry. In the twenty years following the exposition of the principles of reinforced concrete design by Earnest L. Ransome in 1885, the production of Portland cement in the United States rose from 150,000 barrels to 36,000,000 barrels per year. In the years following 1897 when the building trades were temporarily hampered by a shortage in structural steel, the use of reinforced concrete had gained such a popularity that the output of Portland cement was raised from 900,000 barrels in 1895 to 8,400,000 barrels in 1900.

This question has already been fully discussed in another volume.

ADVANTAGES OF REINFORCED CONCRETE

Reinforced concrete possesses many advantages that other building materials do not, and

these have led to its rapid growth as a standard
for many types of structures. The chief advantages that reinforced concrete has are as follows:

Its moderate cost of construction—less than that of
steel, and only slightly greater than that of wood.

Its remarkable fire-resisting qualities, that have been
shown in many instances.

Its strength and its capacity for resisting shock, due
to the monolithic or one-piece nature of the structure.

Its freedom from rotting, to which wood structures
are subject in course of time.

Protection afforded by the concrete to the reinforcing
steel, which would corrode rapidly if left exposed.

Its capacity for resisting the action of many chemical
compounds that would soon destroy structures made of
wood or steel.

Reinforced concrete's low first cost has led
to its use in preference to masonry and steel construction. While wood structures are cheaper
than concrete, the latter are to be preferred on
account of their superior fireproof qualities and
their freedom from decay caused by rot and the
attacks of vermin and insects. Fire insurance
rates for reinforced concrete buildings are only
about half those for wooden buildings of the type
known as "slow-burning mill construction."
The cost of repairs is much less, and no painting
is required to preserve concrete structures subjected to ordinary usage.

Concrete structures may be erected with a
rapidity and ease that are astonishing. Entire
buildings have been erected in the time ordinarily taken to design and form into a whole the

structural metal work for a similar building in steel.

Concrete has the power to resist the action of many chemical compounds that would cause the ultimate destruction of either wood or steel structures. The failure of wood construction is caused by the decay of the timber when exposed to the action of air combined with moisture or chemical acids. In order to preserve wood from this action, it must be covered with some resisting substance, such as sheet lead. This is expensive on account of the high first cost, and the cost of the lead burning required to make tight joints. This form of construction is not entirely satisfactory, as the lead, when exposed to the action of the gases, becomes brittle and soon cracks, allowing the chemical material to escape or become diluted and mixed with foreign substances.

Steel, when used for tanks or other structures containing chemical compounds, will corrode very rapidly, and must therefore be protected by lead or some other substance that will entirely prevent the steel and the chemical from contact. In many cases this protection may be secured by a covering of concrete from one to three inches thick. If the steel be used to resist tensile stresses only, and the concrete to resist compression, the quantity of steel will be reduced, and the cost of construction with it. The resulting reinforced concrete structure will be equally strong and better able to resist corro-

sion. This method would require no extra steel
to be added to prevent corrosion, as is now common practice in the design of steel structures in
chemical plants.

Gases passing through a chimney from a
steam boiler contain sulphur and other impurities which will act upon a steel stack in contact
with them. Steel stacks as a consequence have
a very short life. The use of reinforced concrete
stacks is rapidly growing, because of their low
first cost and their ability in resisting the action
of the stack gases. An inner shell separated
from the outer wall of the stack by an air space,
should be provided to take care of the expansion
caused by the heat of the stack gases.

It has been demonstrated by numerous experiments that concrete will protect steel from
corrosion. Some tests by Professor C. L. Nor-
ton demonstrating this, were made by embedding specimens of steel, clean, as well as in all
stages of corrosion, in stone and cinder concrete
made with both wet and dry mixtures; and then
these samples were exposed to moisture, carbon
dioxide, and sulphurous gases, for periods of
time varying from one to three months. The
steel, where unprotected, rusted completely
away; but when protected by an inch or more
of sound concrete, was not affected.

From these experiments and those made by
other investigators, and from the examination
of structures, which have been taken apart, it
may be concluded that concrete, when properly

mixed and placed, gives the best protection yet discovered for the steel embedded in it, and that concrete may be safely used in all structures except where the chemical action would affect the concrete itself.

Durability. There is scarcely any class of manufacture which is not now being carried on in a reinforced concrete building. It is adaptable to any weight of loading, to high speed and heavy machinery, as well as to light machine tools, and to almost any style of design.

Recent scientific experiments, as well as actual experience, are favorable to the use of concrete under repeated and vibrating loads.

The use of concrete in brackets for supporting crane-runs is an interesting example of severe application of loading. Several concrete buildings in San Francisco withstood the shock of the earthquake, while those around them of brick and stone and wood were destroyed.

While most materials tend to rust or decay with time, concrete under proper conditions continues to increase in strength for months or even for years.

Concrete expands and contracts with changes of temperature. Its coefficient of expansion—that is, its expansion in a unit-length for each degree of increase in temperature—is almost identical with that of steel, and on this account there is no tendency of the steel to separate from the concrete, and they act together under all conditions. As in building with other materials, provision must be made, in long walls or other surfaces, for the expansion and contraction due to temperature, by placing occasional expansion joints or by adding extra steel. In factories of ordinary size, no special provision need be made, as the regular steel reinforcement will prevent cracking.

Stiffness. A reinforced concrete building really resembles a structure carved out of a single block of solid rock. It is monolithic, or of one piece throughout. The beams and girders are continuous from side to side and from end to end of the building, while even the floor slab itself forms a part of the beams, and the columns are also either coincident with them or else tied to them by their vertical steel rods.

All this accounts for the extraordinary stiffness and solidity of a reinforced concrete structure, and puts it into a different class from timber construction, where positive joints occur over every column; and even from steel construction, in which the deflection is greater.

COST OF REINFORCED CONCRETE

As a general proposition, reinforced concrete is almost invariably the lowest priced fireproof material suitable for factory construction. The cost is nearly always lower than that for brick and tile; and, with lumber at a high price, it is frequently even lower than brick and timber, with the added advantage of durability and fire protection.

In comparing the cost of different building materials, one must bear in mind that the concrete portion of the building is only a part of the total cost. Since the cost of the finish and trim may equal or exceed that of the bare structure, even if the concrete itself cost, say, 10 per cent more than brick and timber, the cost of the building complete may not be 5 per cent greater than with timber interior. The lower insurance rates will partly offset this even if there is no other economical advantage for the fireproof structure.

The exact cost of a building in any case is governed by local conditions. In reinforced concrete, the design, the loading for which it must be adapted, the price of cement, the cost of obtaining suitable sand and broken stone or gravel, the price of lumber for forms, the wages of the laborers and carpenters, are all factors entering into the estimate. Reinforced concrete is largely laid by common labor, so that high rates for skilled laborers affect it less than many other building materials.

As a general proposition, it may be stated that the cost of reinforced concrete factories finished complete with heating, lighting, plumbing, and elevators, but without machinery, may run, under actual conditions, from 8 cents per cubic foot of total volume measured from footings to roof, to 16 cents per cubic foot. The former price may apply where the building is erected simply for factory purposes with uniform floor loading, symmetrical design—permitting the forms to be used over and over again—and with materials at moderate prices. The higher price will usually cover manufacturing buildings, located in restricted districts, and where the appearance both of the exterior and interior must be pleasing. This does not include in either case interior plastering or partitions.

Professor C. Derleth, Jr., says:

"No doubt you have been told that there are many systems for reinforced concrete construction, and some of you may be in doubt as to what type should be used. In general, a building of reinforced concrete, designed irrespective of a distinct type or system, will cost more money than an alternate design recognizing a well-developed method. This difference in cost results, not because a patented system has superior merits theoretically, but because it has been tried—because the materials are manufactured according to system and may be readily obtained; moreover, because a company representing a responsible system may be counted upon to furnish high-grade materials."

Along this same line of thought in regard to the intelligent placing of the steel work in reinforced concrete structures, it seems in place to introduce a statement made by Buel and Hill:

"The compressive resistance of concrete is about ten times its tensile resistance, while steel has about the same strength in tension as in compression. Volume for volume, steel costs about fifty times as much as concrete. For the same sectional areas, steel will support in compression thirty times more load than concrete, and in tension three hundred times the load that concrete will carry. Therefore, for duty under compression only, concrete will carry a given load at six-tenths of the cost required to support it with steel. On the other hand, to support a given load by concrete in tension would cost about six times as much as to support it with steel. If the various members of a structure are so designed that all of the compressive stresses are resisted by concrete, and steel is introduced to resist the tensile stresses, each material will be serving the purpose for which it is the cheapest and best adapted, and one of the principles of economic design will be fulfilled.''

In the cost of plain work such as the construction of reinforced walls and foundations, labor has much to do with the cost. The cost of reinforcement for such walls can be easily computed from the current price of steel bars. This cost is sometimes lowered by using material taken from other sources; but in using such material, care should be taken that it is in good condition both as to quality and strength.

If we look at the side of the question which considers only the price of the raw materials in lessening the cost of a structure, we find that the cost of a cement structure can be lessened with respect to the cost of the cement which it contains in two ways: (a) by employing a cheap quality; or (b) by using a small quantity.

A limit as to quality has been mutually agreed upon by a large majority of cement manufacturers and users by the adoption of a standard specification for cement (see under "Materials and Manufacture of Concrete"). As to a minimum quantity, it may again be pointed out that the proper amount depends upon the characters of the other ingredients, upon the uses to which the completed structure is to be put, and upon the strength or quality required of the mixture under the governing conditions.

The cement is the binder holding together the aggregates which make up the major portion of the concrete. These aggregates may be sand and stone or gravel in varying proportions, depending upon circumstances.

As to the relative costs of different mixtures of the various concrete ingredients, it is evident that the poorer the mixture (that is, the smaller the proportion of cement it contains) the less the cost per cubic yard. However, concrete of least cost may not be profitable under all circumstances.

If we consider the **cost of handling** the ingredients, the following points are worthy of notice: Make as much use of the force of gravity as possible. It is a common practice, on large works, to have high trestles and bins for the storage and handling of concreting materials. From these, the materials move downward by gravity through the mixers and into the conveying devices.

Another point is to have the mixing done as near the point of installation of the concrete as possible, since it is easier and cheaper to handle the dry material than the wet mixture. In the case of one large building with structural steel frame, a small electrically-driven portable mixer was used, and moved about in such a manner that it discharged directly at the point where the concrete was to be placed.

The labor item is another place wherein costs may be reduced—not necessarily by employing cheaper help, but by the economical handling of the men on the work. A few extra men who do nothing but the little odd jobs such as keeping planks in place for the men with the wheelbarrows can often save considerable time and money. The forming of a cycle of operations with each man having his particular part, is an ideal condition.

Where several thousand yards of concrete is to be raised to a considerable height, the use of platform elevators, bucket hoists, or derricks is recommended as a means of reducing the cost. The actual cost per cubic yard of hoisting concrete by each of these methods varies by only a few cents. With regard to other mechanical devices, it is sometimes found that the interest on their cost, together with their depreciation, more than offsets the cost of labor performed by them. In case of a bonus allowed for quick work, these devices are often used to an advantage on account of their saving of time.

Every one at all conversant with the costs of reinforced concrete work knows only too well the disproportionate amount due to centering. In very heavy foundation work, this proportion may not be excessive; but in some buildings of reinforced concrete, the labor cost of installing and removing the wooden falsework, together with the cost of the material itself, has made up 50 per cent of the entire expenditure.

The cost of the centering material itself is heavy, and, when wood is considered, is growing heavier almost month by month. This is due to the constant and rapid increase in the prices of timber and lumber of all kinds, which has recently taken place.

The first plan which comes to mind, looking toward the economizing of centering material, is to make repeated use of each piece. This can readily be accomplished where a building is largely a repetition from floor to floor or from bay to bay. But unless the structure is very large, a considerable time is lost through the necessary delays experienced while waiting for the concrete to set before the centers can be removed. This is the case, however, only where the concrete is handled in a wet state.

As an example of the labor connected with the above-mentioned cost of forms and centering, we quote an answer to an inquiry made as to the cost of labor on building walls consisting of piers two feet square with eight-inch curtain walls between. The reinforcement consisted of an ordinary type of deformed bar:

"The labor cost in concrete form building is a very uncertain figure, as no two men do the same amount of work per day. I have had a squad of carpenters and helpers who would build forms for 24 by 24-inch columns at less than 10 cents per foot height; and, again, I have seen it cost as much as 35 cents per foot. Then, conditions make a wide difference; and I recall an instance where forms cost as much as 70 cents per cubic foot of concrete, as no form could be used over again in stories higher up. The side-wall forms are worth from 6 to 20 cents per square foot for labor; but it is all guesswork, for even the nature of the lumber affects the labor cost data; besides, you must remember that the cost of forms is always the big item in reinforced concrete, and until we have a system of changeable forms it always will be. I recall an instance where a contracting firm underestimated the cost of forms $28,000 on a building which they contracted to build for the sum of $230,000. In fact, their estimate as to cost of forms was only $17,000, and the total cost amounted to $45,000.

"The cost of labor per cubic yard for reinforced work of sizes given will be as follows: First-story-columns, $2.35 per yard; walls, $2.75 per yard. Second-story-columns, $2.66 per yard; walls, $3.00 per yard; and so on up. These figures are based on machine mixing and elevator."

A consideration of the detailed costs of concrete construction will show quite clearly where the temptations lie for slighting the work on the part of inexperienced contractors. As an example, take an eight-story office building which was recently erected in the East. In this building, which was 80 feet by 175 feet in ground plan, the percentage costs of the various items were as follows:

Labor 38   per cent
Cement 15   per cent
Stone and lime........................ 9½ per cent

Steel 21   per cent
Lumber 10   per cent
Power .... 1½ per cent
Miscellaneous, unclassified 5   per cent

It is evident that the large items on this job were the labor, the cement, the steel, and the lumber. The lumber charge is for the material used in forms. Although the actual costs in the case of a factory building would be somewhat different, their relation, as shown by these percentages, would hold substantially true, except that there might be some saving in the amount of cement, steel, or lumber. In a word, that the expense of these items could be varied more easily in this type of construction than in most others.

Table VII shows a detailed consideration of the costs of reinforced concrete construction work. The steel cost given at the end of the table applies only to cost of putting in place and does not cover purchase price. These costs were obtained from a very large corporation engaged exclusively in reinforced concrete work and employing as superintendents and foremen experienced, skilled men. The average contractor handling occasional jobs cannot hope to reach these figures except under very favorable circumstances.

In the case of **concrete piles**—another of the many uses to which this valuable agent has been subjected—we again have the question as to how

TABLE VII

Cost of Various Classes of Concrete Work

KIND OF STRUCTURE	AVERAGE COST OF FORMS PER SQUARE FOOT				AVERAGE COST OF CONCRETE PER CUBIC FOOT						
	Carpenter Labor	Lumber	Nails and Wire	Total	Concrete Labor	General Labor	Cement	Aggregate	Teams and Misc.	Plant	Total
	$	$	$	$	$	$	$	$	$	$	$
Building Walls (Above Grade) Average of 17 Structures.........	.085	.036	.002	.128	.090	.016	.073	.076	.025	.019	.301
Footing and Mass Foundations Average of 10 Structures.........	.057	.034	.002	.093	.045	.007	.071	.077	.007	.021	.229
Foundation Walls Average of 14 Structures.........	.068	.033	.002	.103	.076	.015	.080	.062	.019	.017	.269
Beam Floors of Reinforced Concrete Average of 18 Structures.........	.070	.045	.002	.116	.111	.020	.106	.063	.025	.024	.354
Flat Slab Floors Average of 3 Structures..........	.071	.038	.002	.111	.097	.009	.096	.070	.019	.024	.315
Concrete Columns Average of 9 Structures..........	.082	.036	.001	.130	.096	.027	.085	.049	.021	.023	.301
Concrete Slabs Between Steel Beams Average of 13 Structures.........	.061	.032	.002	.095	.102	.019	.128	.068	.024	.017	.359

Cost of handling steel for reinforcement (Average of 21 structures) was $3.52 per ton.

whose domain they are encroaching to a constantly increasing extent.

One of the chief factors making towards the increased cost of wood piles, is their growing scarcity. This is largely due to the recklessness with which our forests are being yearly depleted. The cost of concrete piles, as compared with that of wood piles, was brought out in a striking manner during the erection of the new buildings of the United States Naval Academy at Annapolis, Md. The original plans called for wood piles; but, as the allotment made for the various buildings had been exceeded, it was found necessary to reduce costs wherever possible. Calculations showed that by using concrete piles a saving of over $27,000, or more than 50 per cent of the cost of wood piles, could be effected. As a result, these piles were selected. The various factors which tended toward the economy resulting from the substitution of concrete piles, are thus stated by Walter R. Harper, inspector in charge of the work: 2,193 wood piles were replaced by 885 concrete piles; 4,542 yards of excavation were reduced to 1,038 yards, saving 2,504 yards; and 3,250 yards of concrete footing were reduced to 986 yards, thus saving 2,264 yards. Shoring and pumping, which would have cost $4,000 had wood piles been used, were entirely eliminated. This indicates, in a measure, the means by which foundation costs were reduced as stated. Furthermore, the permanence of the foundation is beyond question. This would not have been the case had wood piles been used. Table VIII shows a detailed statement of the comparison.

The saving in the cost of foundations by the use of concrete piles was $27,458.18, or more than half the original cost of the foundations as designed with wood piles.

TIME FOR SETTING AND HARDENING

The time to be allowed for the necessary setting and hardening of the concrete before the forms are taken away, plays a very important part in the success of the structure. Mr. Edward Godfrey, in his volume on "Concrete," says:

Typical Floor-Slab Construction Framing between Concrete Beams.

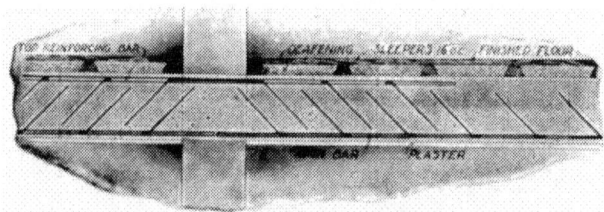

Note Continuous Action over Supports Produced by Inverting Bar in Top of Beam.
DETAILS OF FLOOR-SLAB CONSTRUCTION.

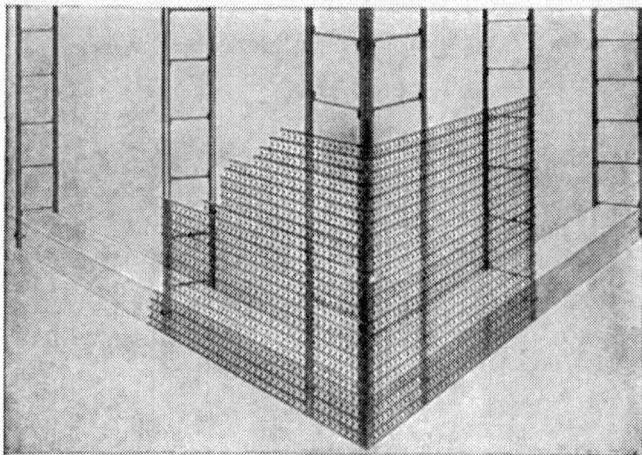

RIB-LATH AND RIB STUDS AS USED IN HOLLOW WALL CON-STRUCTION.

PLATE 15—CEMENT CONSTRUCTION.

TABLE VIII

Comparative Cost of Wood and Concrete Piles

WOOD PILES

2,193 piles, at $9.50..............$20,835.50	
4,542 cubic yards excavation, at $0.40	1,816.80
3,250 cubic yards concrete, at $8.00	26,000.00
5,222 lbs. I-beams, at $0.04.......	208.88
Shoring and pumping............	4,000.00
Total cost	$52,861.18

CONCRETE PILES

855 piles, at $20.00.............$17,100.00	
1,038 cubic yards excavation, at $0.40	415.00
986 cubic yards concrete, at $8.00	7,888.00
Shoring and pumping............	....
Total cost	25,403.00
Difference in cost........................	$27,458.18

"The time that should elapse between the placing of concrete and the removal of the forms depends upon a number of things, among which are the consistency of the concrete, the richness of the mixture, the load sustained, and the temperature and atmospheric humidity. Wet concretes require longer to harden than dry concrete. Lean concretes require longer than rich ones. Concrete hardens more slowly under water or in a saturated atmosphere than in dry air. Low temperatures delay the setting of concrete. If the temperature be below freezing, the setting may be suspended. Failures have resulted on account of forms being removed from concrete which was frozen and which appeared to have hardened through setting.

"Another error apt to be made is to mistake drying for setting. Drying is not a necessary accompaniment to the hardening of concrete. In fact, if the concrete is too warm and the air too dry, the early drying of the concrete that will result will be detrimental to its strength. Concrete should not be allowed to dry out until it has stood for several days.

"Concrete receives its set when it reaches the state where a change of shape cannot be produced without rupture. This requires from a few minutes, in rich mortars of quick-setting cement, to several hours, in lean mixtures. A common way of determining when concrete has set is by pressure of the thumb nail. After the set has taken place, the concrete continues to harden and gain strength for months, and sometimes for years. In ordinary weather, nearly the full strength is attained in six or eight weeks. Loading tests may be made at this stage. Strength necessary to support its own weight is reached at varying periods, depending upon several conditions.

"In counting the time that concrete should stand before removing the forms, those days when the temperature is at or below freezing should be counted out, or at least allowance should be made for almost total suspension of the hardening process during those days.

"It is safe to remove the forms from mass work receiving at the time no load except its own weight, in from one to three days; in warm weather with dry concrete, one day; in cold or wet weather or with wet concrete, more time. When the concrete will bear the pressure of the thumb nail without indentation, it is ready to support itself in this class of work. Thin walls should stand two to five days. Slabs of reinforced concrete should stand about one to two weeks of good weather before being called upon to support their own weight. Slabs of long span may require more time than two weeks. At the same time that the slab centering is removed, or even before it is taken down, the forms on the sides of beams and girders can be removed, leaving the supports of the bottoms in place for a longer time. This will afford an opportunity to inspect the surface of the beams and girders, and to plaster up any cavities before the concrete is too hard. Where practicable, it is well to leave the shores under beams and girders for three or four weeks. Large and heavy beams should be allowed to stand longer than short ones, because the dead weight is a greater fraction of the load they are designed to carry.

"Column forms may be removed in a week or so, if the entire weight of the beams is supported by shores close to the columns; otherwise three weeks or more should be allowed.

"Arches of small span may have the centering removed in one to two weeks. Large arches should harden a month or more. Where practicable, it would be well to leave the concreting of the spandrel wall of an arch span until the arch ring has hardened and the forms are removed. The settling of the arch often cracks the spandrel wall, and gives an unsightly appearance to the bridge.

"Ornamental work should have the forms removed as soon as possible, so that defects can be plastered up, and so that swelling of the wood will have less time to act.

"Falsework should be removed carefully, without jar to the concrete either by hammering on the boards or by dropping heavy pieces on the floor below. The supports should not be removed when any unusual load is on the floor. Materials should not be stored on floors that are not thoroughly hardened and self-supporting.

"Concrete reinforced work should ring when struck with the hammer, before the supports are removed."

FIRE-RESISTING PROPERTIES

Reinforced concrete ranks with the best fireproof materials, and it is this quality, perhaps more than any other, which is responsible for the enormous increase in its use for factories and other industrial structures.

Intense heat injures the surface of the concrete; but concrete is so good a non-conductor that, if sufficiently thick, it provides ample protection for the steel reinforcement, and the interior of the mass is unaffected, even in unusually severe fires.

For efficient fire protection in slabs, under ordinary conditions, the lower surface of the steel rods should be at least $\frac{3}{4}$ inch above the bottom of the slab. In beams, girders, and columns, a thickness of $1\frac{1}{2}$ to $2\frac{1}{2}$ inches of concrete outside of the steel, varying with the size and importance of the member, and the liability to severe treatment, is in general sufficient. In columns, whose size is governed by the loads to be sustained, an excess of sectional area should be provided, as illustrated later.

One of the advantages of concrete construction as a fireproof material is that the design may be adapted to the local conditions. For example, in an isolated machine shop where scarcely any inflammable materials are stored, it is a waste of money to provide a thick mass of concrete simply to resist fire. On the other hand, for a factory or warehouse storing a product capable of producing not merely a hot fire—a hot, short fire will not damage seriously —but an intense heat of long duration, special provision may be made by using an excess area of concrete perhaps two or three inches thick.

Actual fires are the best test of a material. One of the most severe on record occurred in the Pacific Coast Borax Refinery at Bayonne, N. J., and the concrete there, as well as in the Baltimore and San Francisco fires, made an excellent record.

The best fire-resistance materials for concrete are first-class Portland cement with quartz sand and broken trap rock. Limestone aggregate will not stand the heat so well as trap, while the particles of gravel are more easily loosened by extreme heat. Neither of these materials, however, if of good quality, need be rejected for building construction, unless the demands are especially exacting and the liability to fire great. Cinders make a good aggregate for fire resistance, but the concrete made with them is not strong enough for reinforced concrete construction, except in slabs of short span or in partition walls.

The fire resistance of concrete increases with age, as the water held in the pores is taken up chemically and is evaporated.

A recommendation in one of the papers read by Mr. W. M. Bailey before the National Cement Users' Association, stated that if the structural part of the building is of reinforced concrete, not only should the structural steel be protected from fire by covering to a sufficient depth, but also about an inch of additional concrete should be placed on the various members of the structure, to **protect the structural concrete itself**. For example: suppose the design calls for a 12-inch by 12-inch concrete column to carry safely its load. It should be made 14 inches by 14 inches, the extra 1 inch on all faces acting merely as a fire-resisting covering for the concrete inside. In designing buildings of reinforced concrete, it should be borne in mind that concrete is injured by intense heat. When the temperature of concrete reaches about 1,000° F., its surface becomes decomposed, and the water taken up by the cement in hardening is driven off. This process uses a large amount of heat, and is extremely slow after the first ⅜ inch of the outer layer is affected; this outer coat really forms an obstruction through which the heat must pass before each successive layer beyond can be affected. Tests show that a point 2 inches from the surface will stand an outside temperature of 1,500 degrees to 2,000 degrees, with a rise of only 500 to 700 degrees.

Fire Risk and Insurance. When reinforced concrete first came to the front for factories and warehouses, the insurance companies hesitated to assume such buildings as first-class risks. However, examination and tests have gradually convinced the most skeptical of their true fire resistance, until now structures of this material are sought after and given the lowest rates of insurance.

Mr. L. H. Kunhardt, Vice-President and Engineer of one of the oldest of the factory mutual insurance companies—which have for years played a leading part in the development of mill construction and of the science of fire protection engineering and the consequent reduction of fire losses—presents very instructive figures comparing the costs of insurance upon several types of factories for various classes of manufacture. Mr. Kunhardt also indicates the means by which concrete may be utilized in reducing even the present low rates of insurance upon buildings protected by efficient fire apparatus.

From the statements there given, we may conclude that a well-designed reinforced factory with continuous floors (1) offers security against disastrous fires and total loss of structure; (2) reduces danger to contents by preventing the spread of a fire (3) prevents damage by water from story to story; (4) makes sprinklers unnecessary in buildings whose contents are not inflammable; (5) reduces danger of panic and loss of life among employees in case of fire.

Mr. Kunhardt says:

"In consideration of the question of insurance on reinforced concrete factories, the problem simply resolves itself into a determination of what the fire and water damage will be in the event of fire, compared with that in other types of factory buildings.

"For this purpose concrete factories may be divided into two classes:

"(1) Those having contents which are not inflammable or readily combustible. In this class, if wooden window-frames and partitions, etc., have been eliminated,

the building as a whole becomes practically proof against
fire, provided there are no outside exposures, protection
against which would require special precautions.

"(2) Those having contents which are more or less
combustible, and which have in their construction small
amounts of inflammable material, such as wooden window-frames and top floors. In this class the burning of
contents is the cause of damage to the building, the
extent of which is determined by the character of the
contents.

"Of the two, the latter class is the one ordinarily met,
and with which the question of insurance cost is therefore usually concerned. The character of the occupancy,
details of construction, and conditions of various kinds
inside and outside the factory, and in the various communities, have such direct bearing on rates that any
statement as in Table IX of comparative cost must be
extremely approximate, but perhaps of value as showing
somewhat the relative costs. The costs are estimated
upon the basis of a building without a standard fire
equipment, which condition is, however, now rare in the
case of first-class factories and warehouses, even if of
fireproof construction.

"Table IX illustrates in a general way the gain by the
use of the better type of construction; but in factory
work, it has long been recognized that there is a distinct
hazard in the manufacturing operations and inflammable
contents, which is greater in degree than in other classes
of property. The science of fire protection with automatic sprinklers and auxiliary apparatus has therefore
attained such a degree of perfection that the brick or
stone factory with heavy plank and timber floors is obtaining insurance at rates which are lower than those
that are possible on any of the fireproof buildings without sprinklers. The real reason for this lies in the fact
that the contents, including machinery, stock in process,
and finished goods, constitute by far the larger part of

TABLE IX

Concrete Factories vs. those of Wood or Brick

APPROXIMATE YEARLY COST OF INSURANCE PER $100

Exposures, none ; area not large ; good city department ; no private fire apparatus except such as pails and standpipes

CLASS OF STRUCTURE	All Concrete		Brick Mill Construction or Open Joists.		Wood Mill Construction or Open Joists.		Add for brick or wood bldgs. in small towns and cities without best of water and fire departments.
	Bldg.	Contents	Bldg.	Contents	Bldg.	Contents	
General Storehouse	20c.	45c.	60c.	100c.	100c.	125c.	25c.
Wool Storehouse..........	20c.	35c.	40c.	60c.	75c.	100c.	25c.
Office Building............	15c.	30c.	35c.	50c.	100c.	125c.	25c.
Cotton Factory...........	40c.	100c.	100c.	200c.	200c.	300c.	50c.
Tannery...................	20c.	40c.	75c.	100c.	100c.	100c.	25c.
Shoe Factory.............	25c.	80c.	75c.	100c.	150c.	200c.	50c.
Woolen Mill..............	30c.	80c.	75c.	100c.	150c.	200c.	50c.
Machine Shop............	15c.	25c.	50c.	50c	100c.	100c.	25c.
General Mercantile Bldg	55c.	75c.	50c.	100c	100c.	150c.	25c.

NOTE—These costs are based on the absence of automatic sprinklers and other private fire protective appliances of the usual completely equipped building. They are not schedule rates, but may be an approximation to actual costs under favorable conditions based on examples in various parts of the country.

the value of the plant; and these the building alone cannot be expected to protect when a fire occurs within, except in so far as the absence of combustible material in construction may assist in so doing. Fire protection is therefore needed for safety of contents, even if the building itself is practically fireproof.

"As illustrating the value of fire protection, I would state that in the Boston Manufacturers' Mutual Fire Insurance Company, and others of the older of the Factory Mutual companies, the average cost of insurance on the better class of protected factories has now for some years averaged, excluding interest, less than seven cents on each one hundred dollars of risk taken; and on first-class warehouses connected with them, one-half this amount. These figures can be compared with the table as illustrating the gain by the installation of proper safeguards for preventing and extinguishing fire.

"In these same protected factories and warehouses, the **actual fire and water loss** is less than four cents on each one hundred dollars of insurance, and, being so small, it would seem that they must be almost impossible of reduction; but nevertheless it is possible.

"How can this be accomplished? This is the problem of the designer and builder of the concrete factory.

"(1) By avoiding vertical openings through floors— a common fault in many factories with wooden floors. To be a perfect fire cut-off, a floor should be solid from wall to wall, with stairways, elevators, and belts enclosed in vertical fireproof walls having fire-doors.

"(2) By provision for making floors practically waterproof, that water may not cause damage on floors below that on which fire occurs. Scuppers of ample size to carry water from floors to outside are an essential part of the design. In the ordinary factory with wooden floors, loss from water is almost invariably excessive as compared with the loss by actual fire.

"(3) By making the buildings as incombustible as possible, thus reducing the amount of material upon which a fire may feed. Also by provision for sufficient thickness of fireproofing to insulate all steel work thoroughly, the fireproofing being sufficiently substantial that it may not scale off ceilings or columns at a fire or from other causes, thus allowing failure of steel work, by heating or deterioration. An owner is thus more secure if the fire protection or any parts of it fail at a critical moment.

"(4) By good judgment as to the extent or amount of fire protection required in each individual case. While the value of the automatic sprinkler is recognized and the general rules specify its installation, the Factory Mutual companies do not require it in the concrete building, except where there is sufficient inflammable material in the contents to furnish fuel for a fire. An essential feature of good factory construction includes not only consideration of the building, but protection adequate to its needs only.

"The extent to which the above is faithfully carried out, will eventually be the determining feature in the cost of insurance."

G. C. Nimmons, a Chicago architect, has expressed the following views on the relation of concrete construction to fire insurance rates:

"One of the strongest influences toward the increase of the number of concrete buildings nowadays comes from fire insurance companies. The Factory Mutual insurance companies of New England are strong advocates of reinforced concrete buildings for commercial and manufacturing purposes. Not so much for any superior fireproof qualities of reinforced concrete, but on account of the superior waterproof qualities of concrete buildings. I believe it is a matter of record that the Factory Mutuals of New England have paid more for water damage than for fire damage."

General Principles of Rein= forced Concrete Design

The use of some form of steel reinforcement has already been shown to be a necessity. The stresses in a member subjected to bending have been explained, and attention called to the unequal strength of plain concrete in tension and in compression. It remains for us now to find out where to locate the steel reinforcement, and how much is needed for a given load upon a member.

In the case of a concrete beam or girder, the horizontal steel reinforcing rods should be located as near the bottom of the beam as possible, still allowing sufficient thickness of concrete underneath to protect them in case of fire or exposure to liquids, gases, or other agents tending to cause corrosion. This is, as has already been stated, on account of the stresses in the bent beam being divided on each side of an imaginary line located near the center of the beam section (for rectangular beams) and called the **neutral axis.** The stresses above this line are compressive stresses, and are taken care of by the strength of the concrete itself; while those below the line are tensile, and, on account of the weakness of concrete in tension, are taken care of by the steel.

This reinforcing steel may be in the form of

rods or **bars, unit-frames,** or **structural shapes,** for girders, beams, and long spans of construction work, or for short spans bearing heavy loads; but for floors, roofs, etc., of short spans and light loads, the **sheet fabrics** can often be used to advantage.

At the part in the span where the greatest bending action occurs, the depth of the reinforcement below the top surface of a beam or girder varies with designers from $\frac{7}{8}$ to $\frac{10}{11}$ of the depth of the beam section. This allows a good protection against fire in ordinary sizes of beam section, and also allows for plenty of concrete around the metal to resist shearing action along the rods or to prevent the slipping or pulling-out of the rods.

Godfrey, in his work on "Concrete," recommends that the spacing of rods or bars in beams, when the diameter is $\frac{1}{200}$ of the span, should be four diameters for square rods, and three diameters for round rods. If the rods are of smaller diameter than $\frac{1}{200}$ of the span, the spacing may be closer. The distance from center of outside rod to side of beam should be one-half the spacing.

The web of a beam or girder is often protected at the ends against diagonal tension cracks, by **bending up** the end thirds of the rods and bringing their ends nearly to the top surface of the beam. These ends are then fitted with anchors which will hold them firmly in the concrete. This construction has been found to strengthen the web of the beam greatly, and to cause it even to approach the condition of an arch in resisting loads. Professor Talbot found, as a result of many tests, that the loads carried by beams with all the reinforcing bars bent up but not anchored did not differ much from those in cases where the bars were all straight. The failure was slower in the case of the bent bars, and warning of approaching failure was given. He found, however, that if part of the bars were left straight and alternated with the bent ones, the web was considerably strengthened.

These tests also showed that the use of U-shaped stirrups generally strengthened the webs of beams, but the amount of additional strength depended largely upon the quality of the concrete. The stirrups did not exert any considerable strength, however, until a diagonal crack had formed in the beam.

Mr. Ransome's rule for placing the stirrups in a beam is to place the first one at a distance from the end equal to $\frac{1}{4}$ the depth of the beam; the second, a distance of $\frac{1}{2}$ its depth beyond the first; the third, a distance of $\frac{3}{4}$ the depth beyond the second; and the fourth, a distance equal to the depth of the beam beyond the third.

The main point in putting in this steel reinforcement is to get just enough steel below the neutral axis to balance the strength of the concrete above it. We do not wish to use too large sizes of rods in light beams, as the result would be a crushing of the concrete on the top side of the beam, or a shearing along the rods, when the beam was bent. Professor Talbot has shown from his experiments that steel rods whose combined area amounted to from one to one and one-half per cent of the area of the beam section above a line drawn through the center of the rods (the half-holes above this line being figured as a part of the beam section) would allow the beam to fail by tension in the rods. For percentages of steel higher than one and one-half, there is a liability to failure by crushing of the concrete.

These experiments were performed upon beams composed of 1:3:6 concrete. A 1:2:4 concrete would permit of a slightly larger percenage of steel, on account of the greater compressive strength of the concrete.

Tests for adhesion have also shown that the use of rods of a greater diameter than $1/200$ of the length of the beam are liable to pull out of the concrete without breaking. This is due to the want of proper adhesion between the rod and the concrete, in comparison with the strength of the rod itself. Tests have also shown that beams whose depth is greater than $1/10$ the length, or span, when reinforced with horizontal rods, or with rods bent over at the ends and not provided with anchor plates, will fail when loaded to the breaking point, in a manner similar to that shown in Fig. 40. This is due to diagonal tension produced in the beam as a result of the combined vertical and horizontal shear together with the direct tensile stress.

The theoretical discussion of formulas and their deviation will not be discussed in this volume. The theories regarding flexure in reinforced concrete beams are based upon mathematical principles, as well as the principles of mechanics. Therefore the detailed design of any important work—especially when comparative costs must be considered and where failure would result in serious disaster to property or life—should be under the immediate supervision of an engineer trained in these principles and competent to apply them with judgment to the work in hand.

If a study of these theories is desired, reference to Church's "Mechanics of Engineering" will show the so-called **straight line theory;** while the bulletin on "Tests of Reinforced Concrete Beams, Series of 1905," by Professor Arthur N. Talbot of the University of Illinois, gives a demonstration of the **parabolic theory.**

Simple, Practical Rules for Design. Several **empirical formulas,** or working rules based on experience and observation in actual practice as distinguished from refined theoretical calculations, have been suggested for different types of design, and their authors claim that they closely follow the results of reliable tests. These formulas should be used with care and applied only to the class of work that supplied the data from which they were derived.

BARTON SPIDER-WEB SYSTEM IN FLOOR CONSTRUCTION.

PLATE 17—CEMENT CONSTRUCTION.

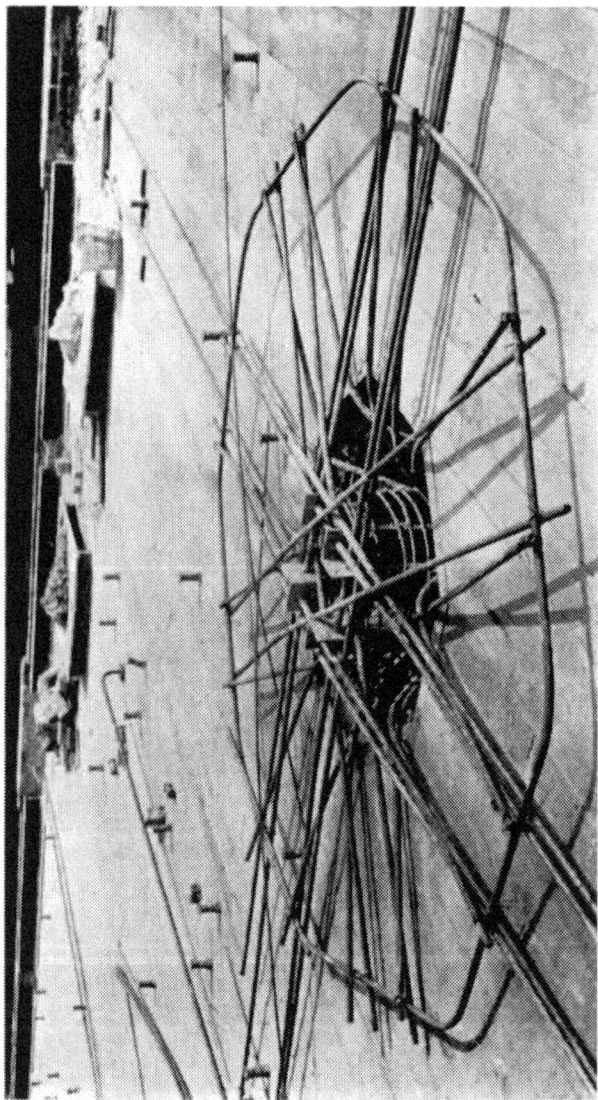

COLUMN HEAD OF SPIDER-WEB SYSTEM.

PLATE 18—CEMENT CONSTRUCTION.

Mr. Homer A. Reid, in his volume on "Concrete and Reinforced Concrete Construction," presents two simple approximate working formulas for the design of beams. These are **Wason's formula** and **Ransome's formula**.

Wason's Formula. This formula is "based on the following assumption: that there is a perfect bond between the steel and the concrete within the limits of the working stresses of the combination. That the steel takes the entire tensile stress and the concrete the entire compressive stress. That the neutral axis is assumed to be half-way between the center of the reinforcing bars and the top of the beam. That the center of pressure of the concrete under compression is considered as being two-thirds of the height from the neutral axis to the top of the beam. The distance from the center of pressure of the concrete in compression to the center of the reinforcement, equals $^5/_6$ of the distance from the top surface of beam to the reinforcement."

$d =$ Effective depth of beam (top to reinforcement).
$1 =$ Span in inches.
$F_s =$ Total stress in steel.
$W =$ Total uniform load in pounds.

Then, taking the center of pressure as the center of moments, the **resisting moment,**

$$M = \tfrac{5}{6} dF_s.$$

The **bending moment** of a freely supported beam under a uniformly distributed load is

$$M = \tfrac{1}{8} Wl.$$

Equating these two moments, and solving for F_s, we obtain:

$$F_s = \frac{Wl}{6\frac{2}{3}d}.$$

Example. Determine amount of steel required for a beam of 12½ ft. span to carry a total uniform load of 12,500 lbs., assuming an effective depth of 14 $^4/_{10}$ inches, and using a unit-stress for the steel of 16,000 lbs. per square inch.

Since 12½ ft. = 150 inches, we have:

$$F_s = \frac{12,500 \times 150}{6\frac{2}{3} \times 14^4/_{10}} = 19,500 \text{ lbs.}$$

$$\text{Area of Steel} = \frac{19,500}{16,000} = 1^{22}/_{100} \text{ sq. in.}$$

Two bars $^{13}/_{16} \times ^{13}/_{16}$-in. give an area of $1^{32}/_{100}$ sq. in.

After determining the total stress in the metal, the area of the reinforcement is determined by dividing the total stress by a safe working stress to determine the area of metal. Bars of proper size are selected to make up this area, a convenient spacing selected, and the area of the concrete adjusted to resist the compression. Mr. Wason uses 16,000 lbs. per square inch tension on the steel; and for a 1:3:6 concrete, an average of 500 lbs. per square inch in compression on the concrete; and requires 32 square inches of concrete in the upper third of the beam for each square inch of steel (in the lower part). This averages very nearly 1 per cent of reinforcement.

The above ratios are applied to the use of
Ransome twisted bars, which have a high elastic
limit and give a factor of safety of about 4.

In the above problem, the total compression
is 19,500 lbs.; this, divided by 500 lbs., gives a
required area in the upper third of the beam
of 39 square inches. $39\times3=117$ square inches;
total area of beam, 117 square inches, divided
by $14^4/_{10}$ inches depth assumed, gives $8^{13}/_{100}$
inches width of beam. A width of $8\frac{1}{4}$ inches
may be used.

Ransome's Formula. Ransome's formula
for a simple beam uniformly loaded is:

$$S = \frac{Wl}{7d}$$

in which,

$W =$ Total dead and live load, in tons.
$l =$ Span, in inches.
$d =$ Depth of steel below top of beam = Effective depth.
$S =$ Maximum stress in beam, either tension or com-
pression.

When the beam is not uniformly loaded, the
formula becomes:

$$S = \frac{BM \times 8}{7d},$$

in which **BM** equals the maximum bending
moment in inch-tons.

In order that the compressive stress per
linear foot of width resulting from a chosen
value of **d** shall not exceed the safe compressive
strength of the concrete, there must be 16

square inches of concrete above the bars for each ton of stress.

$$16S = 12d,$$

from which,

$$S = \tfrac{3}{4}d.$$

Substituting this value of S in the above formula, we have:

$$\tfrac{3}{4}d = \frac{Wl}{7d}$$

$$d = \sqrt{\frac{4}{21}Wl}$$

Having obtained **d,** the total stress in tons, $S = \tfrac{3}{4}$ d.

Example. Assume a flat floor slab, having a span of 12 feet carrying a live load of 150 lbs. per square foot.

It is necessary to assume the dead weight of the floor. Let this be taken as 75 lbs. per sq. ft., making a total load of 225 lb. per sq. ft. The total load **W** in tons on a strip of floor 1 ft. wide would be:

$$\frac{12 \times 225}{2,000} = 1^{35}/_{100} \text{ tons,}$$

and we have for **d,**

$$d = \sqrt{\frac{4 \times 1^{35}/_{100} \times 12 \times 12}{21}} = 6^{8}/_{100} \text{ inches.}$$

The total stress in the bars would equal $\tfrac{3}{4} \times 6^{8}/_{100} = 4^{56}/_{100}$ tons. Assuming an allowable working stress on the metal of 8 tons per sq. in.,

there will be $^{57}/_{100}$ sq. in. of metal required—or
four rods $\frac{3}{8}$ in. square—in each foot width of
slab. When $\frac{1}{4}$-in. rods are used, the distance
from center of rod to bottom should be at least
$\frac{1}{2}$ in.; and $\frac{3}{4}$ in. for $\frac{1}{2}$ in. square rods. For
$\frac{3}{8}$-in. rod reinforcement, we will have a total
thickness of $6\frac{3}{4}$ inches.

In calculating the beam dimensions and
amount of reinforcement for ribbed slabs, the
formula

$$S = \frac{Wl}{7d}$$

is used. This condition, however, is imposed,
that the upper third of the beam, including the
flat slab connecting the ribs, shall contain at
least 5 sq. in. of concrete for each ton of stress
given by the formula. This condition prevents
the concrete in the top of the slab from being
strained beyond its safe compressive strength.

In the design of an ordinary beam which is
to rest freely upon a support at each end, we
have only to consider tension as occurring
in the material on the under side of the
beam. This is not the condition generally
found in reinforced structures. The beams,
girders, and floors, commonly form a continuous
mass, thereby fixing the beams and girders
more or less firmly at the ends. This condition
prevents the ends of the beams and girders
from inclining, as the bending occurs in the center, thereby causing the **top surface** of the beam

to come into tension **over** the supporting walls
or columns. Reinforcing rods must be placed
near the **top** surface of such beams and girders,
and anchored firmly, just as they are placed near
the bottom surface in the middle of the span.

In floor construction where some form of
wire fabric is used as a reinforcing agent, the
sag of the fabric as it is stretched continuously
across several spans brings it toward the lower
surface of the floor slab in the center, thereby
giving it reinforcement in the proper place; and
the fabric slants upward again toward the ends
of the span in order to pass over the beams into
the next span. This upward slope of the material again places strength where it is needed.

When the ends of rods are to be joined in a
continuous construction, unless they are joined
by some form of connection, practice shows that
they should overlap in the concrete for about
fifty diameters.

In the case of **T-beams** of reinforced concrete, the assumption upon which calculations
are often based is that we may substitute for the
actual T-section the area of the rectangle found
by extending the sides of the flange section
downward until they meet a horizontal line
passing through the center of gravity of the end
sections of the reinforcing rods in the lower
part of the web, and we may then figure as in
the case of beams of rectangular section. There
is claimed to be but small error in this
assumption.

The Trussed Steel Concrete Company produce in one of their publications a few interesting phases of stresses in beams, and the action
of reinforcing agents. Fig. 44 shows a beam reinforced with one of their products, in which it

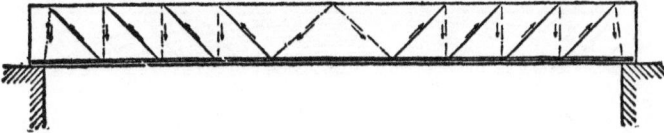

Fig. 44. Truss Action in Beam Reinforced with Kahn Bar.

is claimed there is the action of a complete Pratt
truss. The reinforcing bars should extend to
the top surface of the beam for such an action.

Fig. 45 shows their conception of the truss
action in a beam with horizontal reinforcement
and stirrups. In each of these, the steel diagonals and stirrups form the tension web mem-

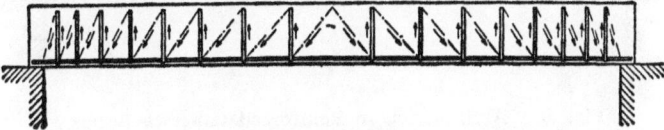

Fig. 45. Truss Action in Beam Reinforced with Horizontal Rods
and Stirrups.

bers, while the compression web members are
supplied by the concrete.

Fig. 46 shows the arch action in reinforced
beams, **A** showing that of a beam reinforced with
Kahn trussed bars; **B**, that of a beam with horizontal reinforcement and stirrups; and **C**, that
of a beam with horizontal reinforcement only.

The principles which govern the **design of reinforced columns** vary somewhat from those

used in the design of beams and floors. Shear plays a prominent part in the failure of a concrete column. When a plain concrete column fails, the general tendency is one of bulging, and sliding of the concrete in the bulged part.

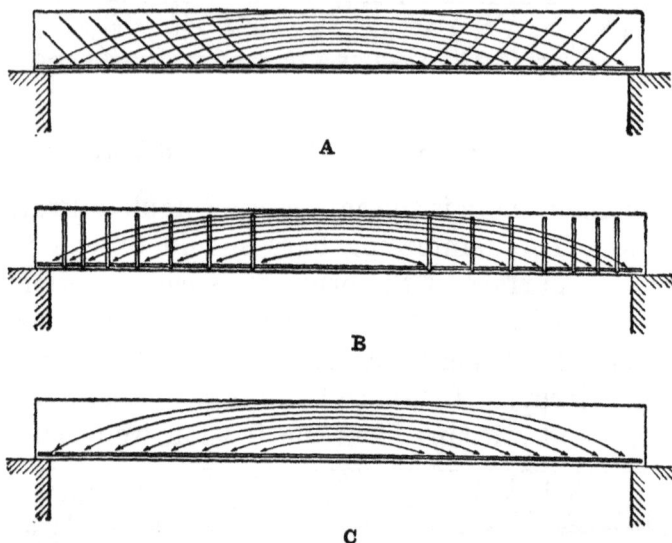

Fig. 46. Arch Action in Reinforced Concrete Beams.

Longitudinal rods of small diameter, when placed near the surface and not connected, do very little to prevent this action, on account of the bending of the rods. Longitudinal rods tied or bound together by bent bars or heavy sheet fabric so as to form a stiff cage surrounding the core of the column, will prevent this buckling action. Many engineers prefer a round or square bar construction to the flat hoop form, on account of better bond in the concrete, especially if the space between the flat bars is small.

We have already referred to the possible tendency in the column to bend when subjected to eccentric loading. If the center of gravity of the applied load at the top of the column does not exactly coincide with the center line of the column, the compression increases on one side of the column and grows less on the other. The limiting boundary within which the center of gravity of the applied load can act, and yet not put one side of the column into tension, is the middle third for a square-section column, and the middle fourth for a round-section column.

Besides the weakening effect of the tension in such a case of loading, the shear (another weakness in concrete) is increased on the compression side of the column. When longitudinal bars bound together as just described are used, the flexural stresses and the tendency to bulge are resisted.

Godfrey suggests that a round or octagonal column with a coil of square steel rods $\frac{1}{40}$ the diameter of the column, the coil itself $\frac{7}{8}$ the diameter of the column, bound together by wiring to the inside of the coil eight longitudinal rods of same size material as the coil, will produce good results. For lengths of columns up to ten diameters, 550 lbs. to the square inch is recommended as the **allowable unit compression** in the concrete. For lengths between ten and twenty-five diameters, he suggests that the

TABLE X

Materials Used for Floors and Roofs—Weights per Sq. Ft.

When slabs are covered with any of the following materials the weight is considered as forming part of the superimposed safe

MATERIALS.	FLOORS	WEIGHT IN LBS. PER SQ. FT.
7/8″, Single thickness flooring, wood..............		3.00
2″x4″ Spruce sleepers 16″ ctrs., and 2″ dry cinder concrete filling		8.50
Asbestone flooring 1/2″ thick....................		3.50
Rubber tiling ....1.00 to		3.00
Tiling .... 3.00 to		8.00

CEILINGS

3/4″ Wood ceiling................................	2.50
Corrugated iron	1.00
Stamped steel	2.00
Metal lath and plaster.........................	10.00
6″ Hollow tile.................................	23.00
8″ Hollow tile.................................	28.00
Plastering	5.00

ROOFS

Common shingles	2.50
18″ Shingles	3.00
Slate, 3/16″ thick..............................	7.25
Slate, 1/4″ thick..............................	9.60
Plain tile or clay shingle.................11.00 to	14.00
Ludowici tile	8.00
Copper sheets	1.50
Tin, including one thickness of felt.................	1.00
Five-ply felt and gravel.........................	6.00
Four-ply felt and gravel.........................	5.50
Three-ply ready roofing........................	1.00
Skylights with galvanized iron frame.............	5.00
Sheathing, 1″ thick, Pine or Hemlock.............	3.00
Sheathing, 1″ thick, Yellow Pine........	4.00
2″ Book tile..............................	12.00
3″ Book tile..................................	20.00
8 1/2″ Solid tile..........................	16.00

TABLE XI

Allowable Floor Loads in Accordance with the Building Laws of Various Cities

LIVE LOADS FOR FLOORS IN DIFFERENT CLASSES OF BUILDINGS EXCLUSIVE OF THE WEIGHT OF THE MATERIALS OF CONSTRUCTION.	New York 1902	Chicago 1902	Phila-delphia 1902	Boston 1902	San Francisco 1906
	Pounds per Square Foot				
Dwellings, Apartment Houses, Hotels, Tenement Houses or Lodging Houses.	60	40	70	50	60
Office Buildings, (1st floor).............	150	100	100	100	150
Office Buildings, (above 1st floor).......	75	100	100	100	75
Schools or Places of Instruction........	75			80	75
Stables or Carriage Houses..............	75	40* 100†			75
Buildings for Public Assembly.........	90	100	120	150	125
Buildings for ordinary stores, light manufacturing, and light storage.....	120	100	120		120
Stores for Heavy Materials, Warehouses and Factories.........................	150		150	250	250
Roofs—(Pitch less than 20 degrees).....	50	25	30	25¶	50
Roofs—(Pitch more than 20 degrees)....	30	25	30	25¶	30
Sidewalks................................	300				300
Public Buildings except Schools........				150	

* Stables less than 500 square feet in area.
† Stables over 500 square feet in area.
¶ Make proper allowance for wind at 30 lbs. per square foot horizontal.

formula

$$p = 670 - 12\frac{l}{D},$$

in which l is the length of the column in inches and **D** is its diameter in inches, will give the **allowable pressure** per square inch in the section. Twenty-five diameters should be the limiting length of column.

In columns which carry heavy loads, as in

TABLE XII
Building Laws Regarding Reinforced Concrete

ITEM	NEW YORK	CLEVE-LAND	SAN FRANCISCO	BUFFALO	TORONTO	PRUSSIAN REQUIRE-MENTS
Ratio of Modulus of Elasticity of Steel to Concrete...........	12	15	15	12	12	
Tensile Stress in Steel...........	16,000	16,000	$\frac{1}{3}$ EL	16,000	16,000	17,000
Compressive Stress in Steel.....		12,000			12,000	
Shearing Stress in Steel.........	10,000	10,000	10,000	10,000	10,000	
Extreme Fiber Stress on Concrete in Compression..........	500	* 500	500	500	500	$\frac{1}{5}$ U
Concrete in Direct Compression.	350	* 400	† 450	350	350	$\frac{1}{10}$ U
Tensile Stress in Concrete.......	0	0	0	0	0	0
Shearing Stress in Concrete.....	50	50	75	50	50	64
Bending Moment in Beam Continuous	$\frac{1}{8}$ wl	$\frac{1}{10}$ wl	$\frac{1}{8}$ wl	$\frac{1}{8}$ wl	$\frac{1}{8}$ wl	$\frac{1}{10}$ wl
Bending Moment in Slabs Continuous	$\frac{1}{10}$ wl	$\frac{1}{10}$ wl	$\frac{1}{12}$ wl	$\frac{1}{10}$ wl	$\frac{1}{10}$ wl	$\frac{1}{10}$ wl
Bending Moment in Square Floor Panels..................	$\frac{1}{20}$ wl	$\frac{1}{15}$ wl	$\frac{1}{20}$ wl	$\frac{1}{20}$ wl	$\frac{1}{20}$ wl	
Method of Calculation...........	S. L.	S. L.	S. L.	S. L.	S. L.	S. L
T-Section Amount Allowed as part of Beam..................	10 b	6 b	5 t	10 b	5 b	$\frac{1}{3}$ l
Columns - Maximum Ratio of Height to Width	12	16	15	16	12	18
Requirements of Tests .	3 L	3 L	2 L	3 L	3 L	2 L + D

NOTE—S. L.=Straight Line Formula
    b=Breadth of Beam   t=Thickness of Slab
    w=Total Uniformly Distributed Load
    l=Length of Beam
    U=Ultimate L=Live Load D=Dead Load
    E L=Elastic Limit
    † Hooped Columns—700 lb
    * 600 lb. If tests show factor of safety of 5 in ninety days.

high buildings, a more suitable construction is that of a steel framework, built up of channels or angles and lattice-work, which is then filled with and surrounded by concrete.

Other valuable formulas and rules for design will be found below, under "General Building Construction," in connection with the Report of Committee on Reinforced Concrete of the National Association of Cement Users. Tables X, XI, and XII contain much data that will be found of practical value in the design of concrete and other structures.

Reinforcing Materials and Systems

Steel is the common medium used in reinforcing concrete. Although the ways and forms in which it is used are varied, and allowing that each form may have its special advantage, the general principle of placing the metal where it will take up the tensile stresses is common to all forms.

Some of the more common methods of application are as follows: by steel bars, rolled shapes, sheet metal, expanded metal, woven

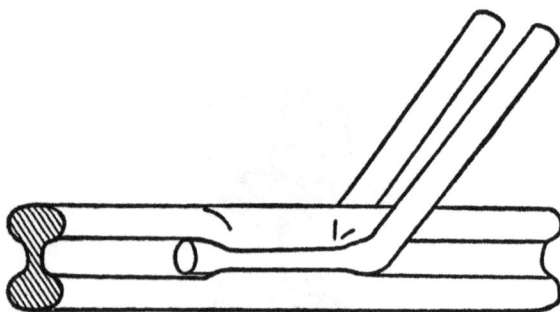

Fig. 47. Monolith Steel Bar.

wire, plain and barbed wire, old structural work of various kinds, wire rope, chain, etc.

Steel bars are divided into different classes —**plain bars, deformed bars,** and **trussed bars;** Plain bars, comprise the ordinary round, square, and flat bars; while deformed bars are illus-

trated by the "Johnson" bar, "Diamond" bar
and "Ransome" bar.

An example of the trussed form is shown in
the "Kahn" bar. These forms and many
others are shown in Plate 12 and Figs. 47 to 51.

The round bar is generally preferred to
either a square or flat bar, on account of the
liability of cracks in the concrete developing in
setting when the sharp-cornered bars are used.

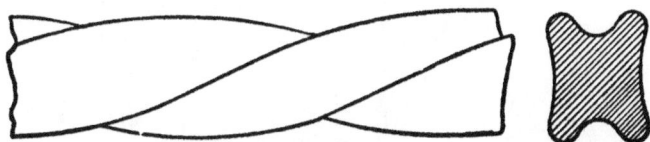

Fig. 48.   U-Twisted Lock Bar.

The round bar also has the advantage of being
easily threaded and fitted with nuts at the ends,
in case plates for prevention of slipping of rod
in the concrete are to be used at the ends of the
bar.

In bars like the Kahn trussed bar and others
of a similar type, the diagonal arms prevent the

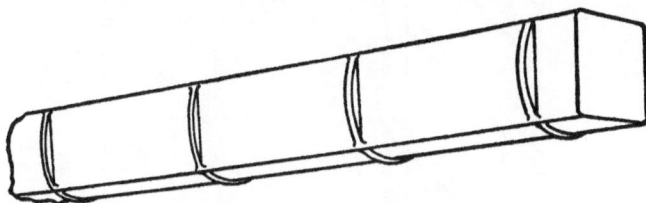

Fig. 49.   Square Lug Bar.

lower main part of the bar from slipping, and
strengthen the beam or girder against the effects

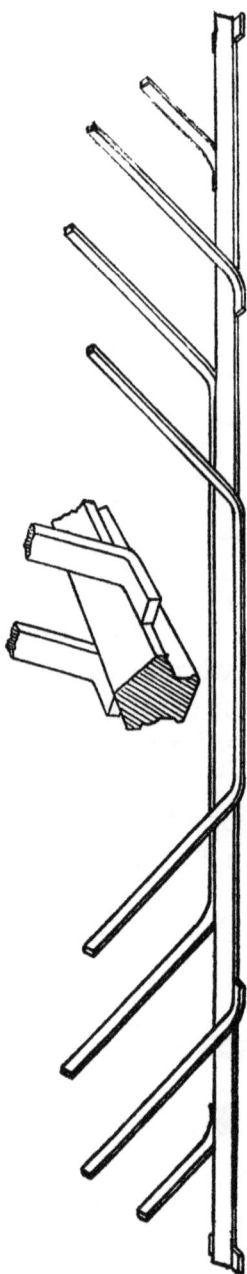

Fig. 50. Kahn Trussed Bar.

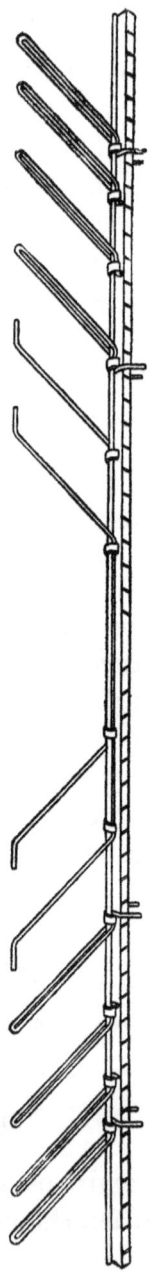

Fig. 51. Herringbone Trussed Bar.

COLUMN IN TURNER "MUSHROOM" SYSTEM.

PLATE 19—CEMENT CONSTRUCTION.

STEEL BEAM FOR REINFORCED CONCRETE CONSTRUCTION
LOADED FOR TEST.

of internal stresses throughout its depth.   In ordinary forms of deformed bars, the primary object in each is to provide a better bond against slipping between the bar and the concrete. The relative merits of the various bars will not be discussed; but according to tests conducted by Professor Talbot, it is questionable whether there is really any great gain in strength effected by the use of the ordinary deformed bar.

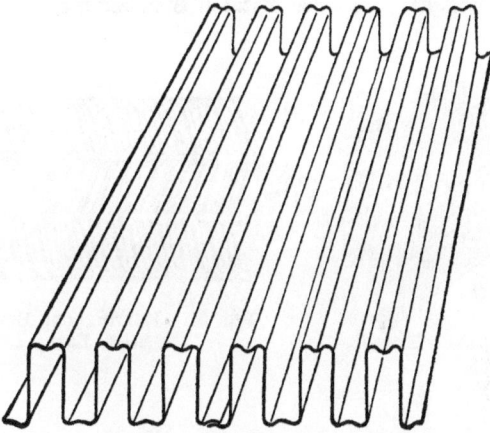

Fig: 52. Multiplex Steel Plate.

Column hoops are formed by bending either flat, square, or round bars into a spiral which is held in shape by longitudinal stays having various means of holding the metal to the stay. An eccentric load placed upon a column tends to throw one side into compression, and the other into tension.   Therefore columns need near-surface reinforcement, just as beams when subjected to similar stresses.

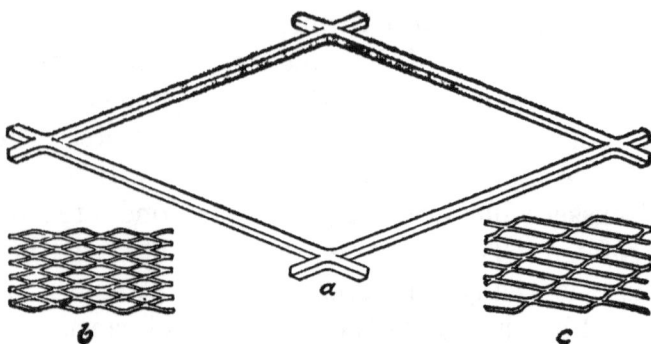

Fig. 53. Expanded Metal of Standard Mesh.

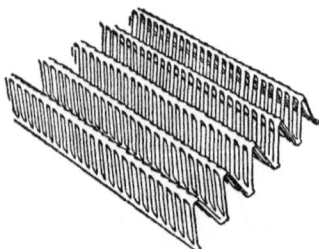

Fig. 54. Trussit.

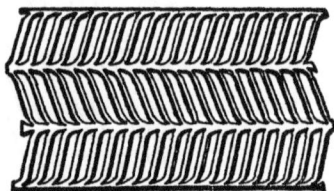

Fig. 55. Herringbone Expanded Steel Lath.

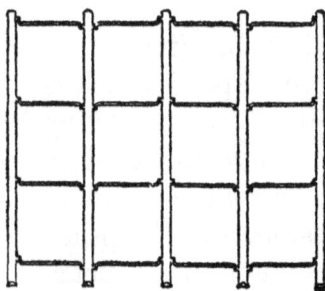

Fig. 56. Rib Metal.

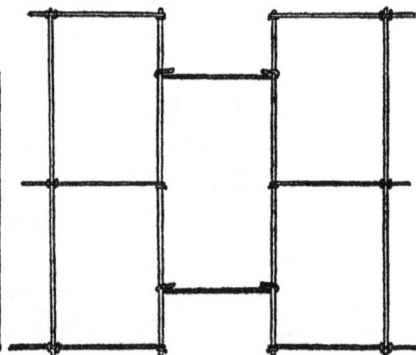

Fig. 57. American System Wire Fabric and Clamp.

Steel shapes, possibly more commonly of the I, T, channel, and angle forms, are used extensively in floors, girders, columns, and even in heavier foundation work.

Sheet metal, expanded metal, metal lath, woven wire, etc., are used in roofs, ceilings, sidewalks, concrete piles, partitions, short-span floors, columns, ornamental work, or, in fact, in almost any construction where there are not too great stresses, or where flexibility within the

A   B

Fig. 58. Lock-Woven Fabric of Standard Gauge.

A—Long wires, No. 10 gauge, on 4-in. centers; cross wires, No. 9 gauge, on 6-in. centers. B—Long wires, No. 3 gauge, on 4-in. centers; cross wires, No. 10 gauge, on 6-in. centers.

forms is needed. Plate 13 and Figs. 52 to 58 show some of the various forms of these materials.

It is not uncommon in light work to find ordinary **strong wire,** and often **barbed wire,** a very valuable reinforcing agent. In rough foundation work, **scrap iron** is often used as a bonding material; but the use of cast iron, whose tensile strength is exceedingly low, would hardly be recommended in any place where bending might occur. Broken **machine parts** and old **iron hoops** from barrels, are often used

on the farm in rough construction work. Old
steel rails are very conveniently used in footings, but they should be clean and free from oil
or heavy rust. In fact this requirement should
apply to all material which is to be embedded in
concrete. A little rust is sometimes beneficial
in helping to remove the mill scale from the new
bars, thus allowing the concrete to obtain a
better hold.

Wire rope or **cable** is an unsatisfactory form
of reinforcement, not only on account of its flexibility, but also on account of the stretch resulting when under stress. Tests have shown a
wire cable to stretch four times as much as a
plain steel rod under the same unit-stress.
Therefore a beam reinforced in this manner
would probably crack under low stresses.

Mr. Edward Godfrey, in his volume on "Concrete," speaking in regard to the quality of the
steel used in reinforcing, touches upon some
vital and debatable points on which there has
been much controversy, when he says:

"Steel for reinforced concrete should preferably be
open-hearth steel, though Bessemer steel may safely be
used for rods and for plates and shapes that are punched,
if the punched holes are reamed.

"The ultimate strength of the steel is not a matter of
much importance, neither is the elastic limit, except as
these properties indicate uniformity in the product. It
should be a good grade of soft steel. It is of more importance that it stand the bend test of soft steel than
that its ultimate strength and elastic limit be high. The
reason why high elastic limit and high ultimate strength

are not essential characteristics of steel embedded in concrete, is the simple fact that these qualities cannot be made use of in proper design of reinforced concrete. This is directly contrary to a great amount of trade literature and some technical literature. Commercial soft steel is almost universally of an ultimate tensile strength of from 50,000 to 60,000 lbs. per sq. in., and a strength at elastic limit of 30,000 to 40,000 lbs. per sq. in. This latter is about three times the safe value that ought to be allowed on the steel, because above this value cracks begin to appear, and there can be no justification for a design that anticipates cracked beams and slabs. There is therefore ample margin of safety in any good soft steel.

"If high steels showed smaller elongations for a given unit-stress (for stress within the elastic limit) than soft steels, there would be some justification for their use, as they would then not stretch out as much under a given stress, and the concrete would be less liable to be cracked. But the modulus of elasticity is one property that is practically constant for all grades of steel. Even soft wrought iron has a modulus of elasticity almost as great as the hardest steel. The simplest conception of the modulus of elasticity, designated as **E,** is a unit-stress that would stretch a piece of steel out to double its original length, at the rate at which it stretches within the elastic limit. The modulus of elasticity of steel is about 30,000,000; if, then, a piece of steel is stressed to 10,000 lbs. per sq. in., it will stretch one-three-thousandth of its length. Beyond the elastic limit, different grades of steel exhibit different characteristics. Soft steels stretch out more before failure, while high steels and soft steels that have been rolled or drawn cold or twisted cold, break without much stretch or reduction of area at point of fracture. This lack of stretch beyond the elastic limit is held out as a benefit in trade literature. It is a positive detriment. If failure occurs in steel that will not stretch, it will be sudden and without warning; whereas, if the steel stretches out, it will allow a beam or slab to sag before failure. Besides giving warning of failure, the sagging will in many cases reduce the stress in the steel very materially. The author has seen tests of slabs reinforced with soft steel that sagged enormously and could not be broken.

"The reason why it is important that steel shall stand the cold-bend test is because rods are very often curved and bent in construction. This bending should be done cold; for, if the steel is heated, its internal structure is changed, and annealing would be necessary to restore it. Soft steel of ordinary manufacture will, in general, stand more punishment than harder grades of steel. The threading of rods and punching of plates or shapes are less liable to cause incipient cracks or hardened metal in soft steel than in high steel. These are also processes to which the embedded steel may be subjected.

"Special steel, while it has a high sound, does not possess any needful characteristics, as an element in reinforced concrete, that are not possessed by the cheap commercial article. This is true because of the limitations of the concrete. Good soft steel is not a special steel, but is the commonest product of the steel furnace. It is important, however, that it be good and that it be soft—that is, not a high-carbon steel.

"There should be a wide margin of safety in the amount that a steel rod will bend. A piece of steel of high ultimate strength may stand a bend of 100 degrees, and fail if bent 105 degrees. It is clear that this steel would not be fit to use where it is bent at an angle anywhere near approaching this.

"The best carbon steel for structural purposes is found to possess an ultimate strength between 55,000 and 65,000 lbs. per sq. in. The elastic limit of the steel should be not less than one-half of the ultimate strength, and the stretch in a measured length of 8 inches should be not less than 24 per cent. It should stand the cold and quench-bend test, 180 degrees flat, without fracture. In the quench-bend test steel is heated to a cherry red, as seen in the dark, and quenched in water at ordinary temperature, before bending.

"The proper use of steel in concrete is in small sections well distributed throughout the mass. The bars should be separated from each other to give the gripping effect of the concrete full play. If they are placed in a layer close together, a cleavage joint is formed, and the concrete is liable to break off. This would be the case in a close coil or set of rings close together or in a set of rods lying close together near the bottom of a beam. Plates or sheets of steel should not be used as separators or spacers for rods, as these will also form cleavage planes. If heavy rods are used, surrounded by comparatively little concrete, the concrete is unable to grip the steel, and the differential expansion due to change in temperature will crack the concrete. Pyramids of concrete surrounding the bases of outdoor steel columns seldom last through a winter without being cracked. Girders covered with a shell of concrete would be subject to the same detrimental effect of differential expansion.

"The steel should not be placed close to the surface of the concrete. It cannot be gripped properly unless it is deep enough in the concrete for the latter to take hold. It cannot be protected from rust and fire unless there is some concrete between the steel and these destroying elements. It is bad practice to lay the steel on the forms, and then the concrete on this. The steel is neither properly protected nor gripped by such means. The depth to which steel should be buried depends upon the size of the section. It is manifest that the heavier the section, the more concrete is needed to grip it and to overcome differential expansion. If rods are bedded deeper they will be less affected by external change of temperature. Heavy rods are of more importance in a structure, hence their protection is of more vital importance than that of light rods.

"Standard sizes of rods or shapes should be used as much as possible, so that they can be obtained without delay from the mills. Also, as few different sizes as possible should be employed. Simple details are essential. A complex structure will be difficult to surround properly with concrete. There should be no broad, flat surfaces to work the concrete under.

"The steel work should be designed with a view of its being easily placed in proper position, and held there against the displacing tendencies due to the placing of the concrete. Where rods cross, they should be wired together, and this should be done before the forms are erected to the point that they will interfere by preventing free access to the rods. Extra wires may often be used to advantage, to tie the rods in place. These wires may serve the further purpose of holding the sides of the forms from spreading. They can be cut off at the surface when the forms are removed.

"Rods in the bottom of a slab can be kept from lying on the bottom of the form by placing small stones under them just before the concrete is placed."

Bars for reinforcing may be **bent** or **twisted** either at the work or at the manufacturing plant. Twisted bars can be readily obtained from the manufacturers; but in cases of large contracts, the twisting is often done at the place of installation. The number of twists per linear foot depends upon the diameter of the bar; thus, for $\frac{1}{4}$-inch bars, there may be five twists per foot; and for 1-inch bars, one twist per foot. In computing the cross-sectional area of steel in reinforced concrete, the twisted bars are figured as square bars of the dimension before twisting.

Fig. 59 shows a pair of tongs which may be used in bending light steel bars during the construction work. Fig. 60 shows a power bender used for the heavier shapes. Light bars may be bent cold. In the case of heavier bars which are heated and then bent, care should be taken to see that the steel is properly annealed afterwards, to restore its original properties.

Life of Steel in Concrete. Good concrete is one of the best known preservatives of steel. The mixture should be wet when the steel is

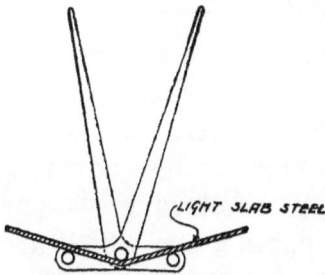

Fig. 59. Tongs for Bending Light Steel Bars.

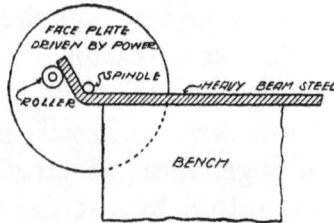

Fig. 60. Power Bender.

embedded, and thoroughly tamped so as to coat the steel completely. Dry or porous mixtures allow access to the steel of air or other vapors, which should be avoided. It is generally assumed that steel embedded in concrete does not corrode. Instances where steel has stood in concrete for years seem to prove this. In fact, it has been stated that steel with a slight coating of rust, when placed in position, has been found to be bright when removed after considerable lapse of time. An exception might be made to the above statements in case of injury to the concrete causing cracks, since rust or corrosion is liable to occur at these cracks.

An objection has been made to the possible corrosion of steel embedded in cinder concrete through the action of sulphur in the cinders; the case, and that the corrosion resulted mainly but experiment seems to show that such is not from the rust, or iron oxide, in the cinders. Cinder concrete should be especially well rammed while wet.

SYSTEMS OF REINFORCEMENT

Although we have referred in a general way to the use of various forms of steel as reinforcing agents, architects and engineers have begun to use these materials in more or less advantageous grouping of parts. The following **systems** illustrate a few of the combinations which are in use in building construction to-day. Most of these are the results of careful computations by competent men, and some have been thoroughly tested by severe conditions in construction work.

The **Ransome System** is one of the pioneers in the line of reinforcing concrete floors and beams by the use of rods. This system consists of reinforced beams, quite thin and deep, with web stiffeners, spaced about like floor joists in ordinary construction work, and supported by girders at the ends. Several bars are grouped in parallel layers for the heavier girder reinforcements. In the case of long spans in the beams, a central stiffening web is often used as bridging between the beams. See Figs. 61 and 62 for details of construction as applied to the Pacific Coast Borax Refinery at Bayonne, N. J.

The **Kahn System** of reinforcing by means of the **Kahn Trussed Bar** is shown in Plates 14

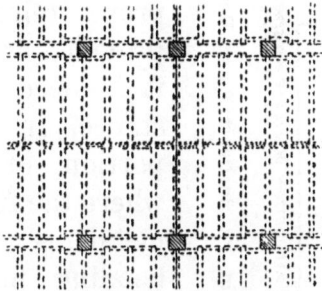

Fig. 61.   Ransome System.
Plan of typical floor.

and 15. The detail of this bar is shown in Fig. 50. This system really consists of what may be considered as a large number of separate members all rigidly attached and handled as a unit. The labor saving in handling a single piece as compared with many separate individual parts, is one of the advantages claimed for this system.

The Kahn trussed bar is made of a special grade of medium open-hearth steel with an elastic limit up to 42,000 pounds and an ultimate tensile strength of 70,000 pounds. The cross section illustrated has two horizontal flanges or wings projecting at diametrically opposite corners. These winged portions are sheared up

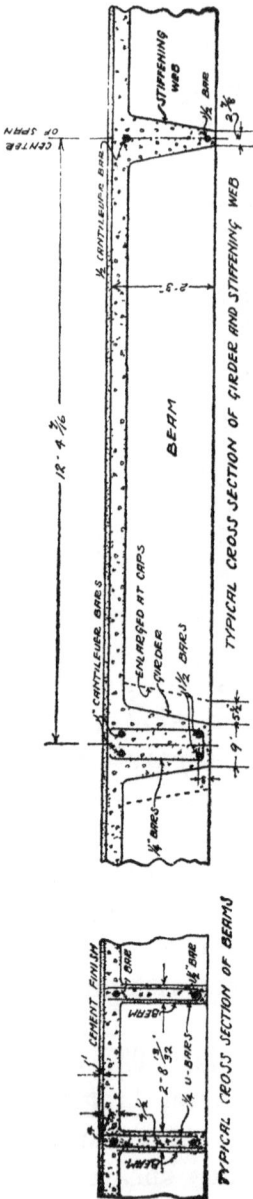

Fig. 62. Ransome System—Cross-Section of Pacific Coast Borax Refinery.

at intervals and bent so as to make an angle of 45 degrees with the main portion of the bar. It is claimed that in this bar there is no waste metal at any point and proper reinforcement is provided at every place it is needed. For instance, in the central portion of the beam where the full section of metal is needed for bending moment, and no reinforcement required for shear, the bar is unsheared and the full area of the metal is available. At the ends of the beam where the shear is a maximum and the bending moment a minimum, the flanges of the bar are struck up to form rigidly attached shear members which carry the shearing stresses directly to the main tension member.

The **Gabriel System** of

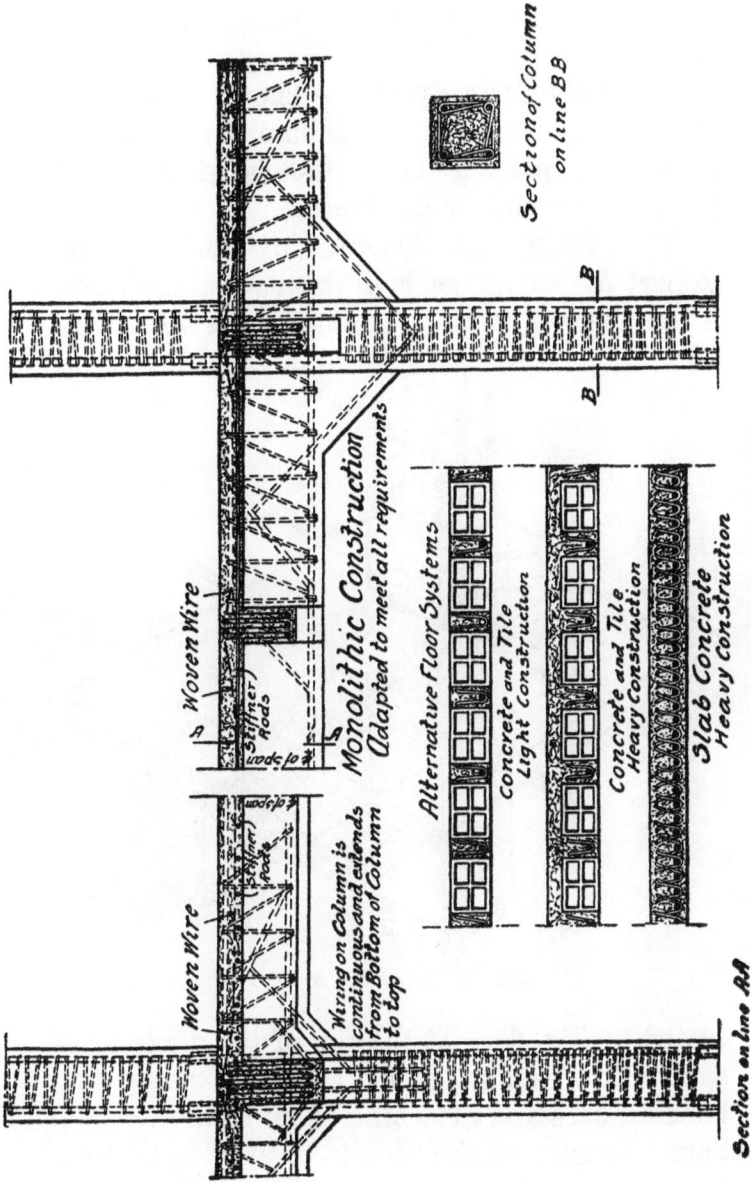

Fig. 63. Gabriel System.

reinforcing beams, floors, and columns, is shown in Fig. 63.

The **Columbian System** is shown in Fig. 64. The special forms of Columbian bars, and the methods of placing them in slabs, girders, and columns, are clearly illustrated.

The **Cummings System,** invented by Mr. Robert A. Cummings, is in this special bar class. A number of reinforcement details are illus-

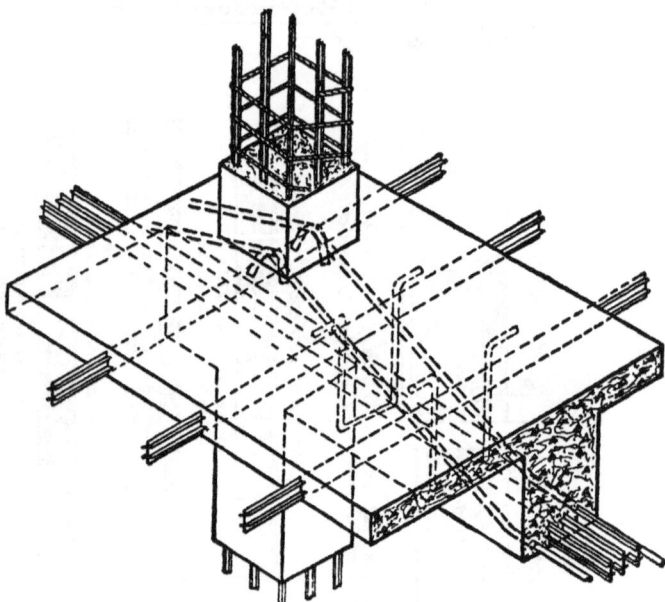

Fig. 64. Columbian System.

trated in Fig. 65. At the top of the diagram is shown the Cummings method of forming the bent-up bars and attaching them to the tension bars. In general the plan is to provide tension bars with ends specially anchored; while se-

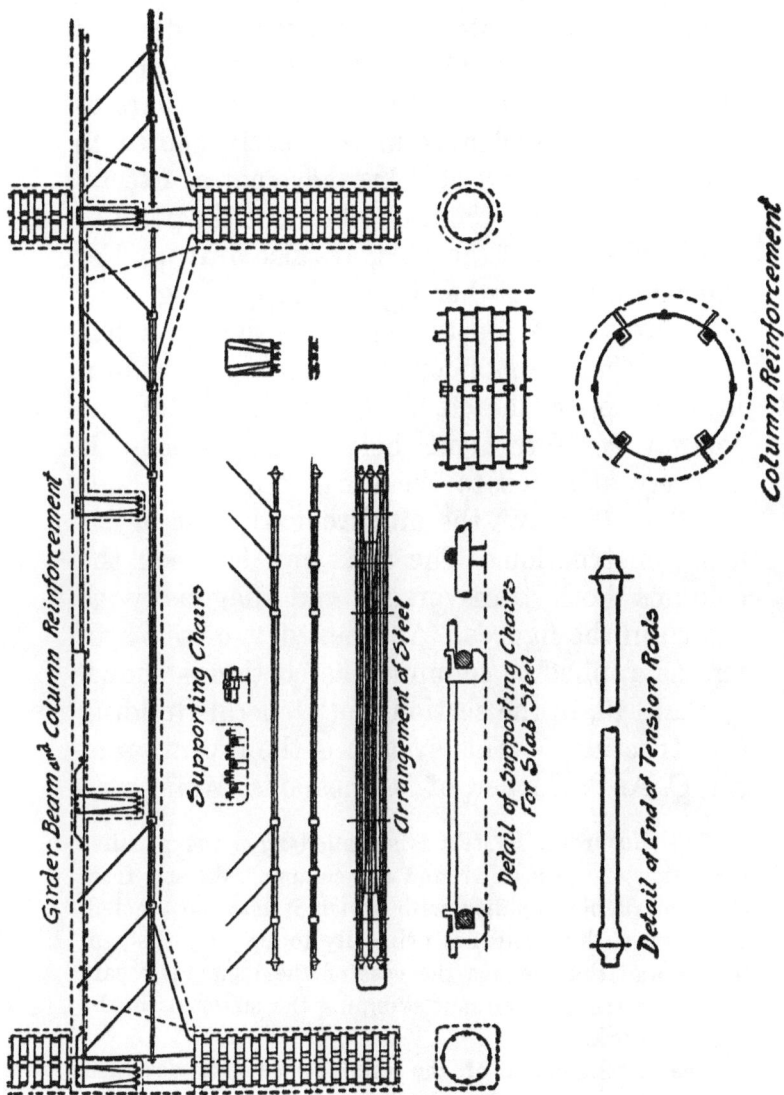

Fig. 65. Cummings System.

curely attached to them are small rods horizontal in the middle of the beam or girder, but
bent up, as indicated, to pass across the top of
the beam and form inclined inverted U-bars or
stirrups. The idea is more clearly shown in
the sketches below of "Arrangement of Steel."
The "Supporting Chairs," placed at the point of
the bending up of the rods, are also drawn. For
the slab steel, another type of supporting chair
is employed, as illustrated in the detail sketch.

The Cummings hooped column is also shown
in the upper sketch, and the details of the
column reinforcement below. Each hoop is
securely attached to the upright rods.

Plate 16 shows the **Mushroom System** of flat
slab construction. The rods run between the
columns both transversely and diagonally as
shown in the figures. Another view of this system as applied to column construction is shown
in Plate 19, in the section on "General Building
Construction." This system is the invention of
Mr. C. A. P. Turner, of Minneapolis, Minn.

The **Mushroom System** is so-called from the peculiar
formation of the rods around the column head and from
the remarkable rapidity with which it may be erected.
The idea of this system is primarily to simplify the centering, and thus reduce the cost of the temporary part
of the construction without skimping the materials in the
finished work.

The arrangement of the reinforcement is designed
with a view to securing the maximum efficiency of the
materials through straining the concrete in a number of
directions, the compression in one direction tending to

THREE-HINGED SEGMENTAL ARCH TRUSSES OF REINFORCED CONCRETE.

Replacing timber viaducts at plant of Armour & Company, Union Stock Yards, Chicago. Arch-ribs spaced 12 ft. apart; rise 15 ft. 6 in.; cross-section of ribs, 20 by 15 in.; reinforced with four 2½ by 2½ by ¼ in. angles latticed together.

NEIGHBORHOOD CLUB HOUSE, SHERMAN PARK, CHICAGO, ILL.

Solid reinforced concrete. A typical example of the field houses that adorn all the small parks of the city.

balance and offset that in another; incidentally to concentrate the maximum amount of reinforcement around and over the support where the shear is the greatest; to enable removal of the forms at the earliest possible period; and finally to eliminate beams and ribs which interfere with light and catch dust, cost money to plaster and finish, and reduce the clear story height. The flat ceiling so obtained gives free and unobstructed illumination from the windows; it permits one to place partitions anywhere without regard to the floor, giving an unusual stiffness and solidity due to the fact that a part of the material which in the beam type is placed in the ribs, is consolidated in the slab, making the slab of unusual thickness with an actual decrease in the total amount of material where the loads are at all heavy.

In fighting a fire where the contents of the building are particularly inflammable, those who have given the matter attention will understand how a rib in the ceiling stops the stream of water which is elevated sufficiently to strike the ceiling, whereas, given a flat ceiling it is easy to reach any part of the floor desired, for, if the stream is elevated and strikes the ceiling, it merely glances along and is spread over the floor where required, instead of being stopped short where it strikes a projecting beam or rib.

The Mushroom System is adaptable to nearly all classes of buildings—court houses, state capitols, warehouses, factories, hotels, bridges, and high office buildings. In the office building partitions can be readily shifted to suit the tenant without interference of rib framing, while cost of plaster finish is but 65 per cent of that where beams are used; and for a given clear story height, there is a saving of 10 per cent of exterior walls.

In a factory building the advantages claimed for this system are:

1. Better distribution of light.

2. The flat ceiling enables more convenient placing of shafting than where ribs are used.

3. Freedom from vibration. Owing to the fact that any concentrated load brings into action the reinforcement of the entire panel, it has little local effect, and consequently the construction is subjected to a minimum of vibration.

The **Barton "Spider-Web" System** of reinforcing floor slabs is shown in Plate 17, while the method of forming column caps is shown in Plate 18. The interesting features of this sys-

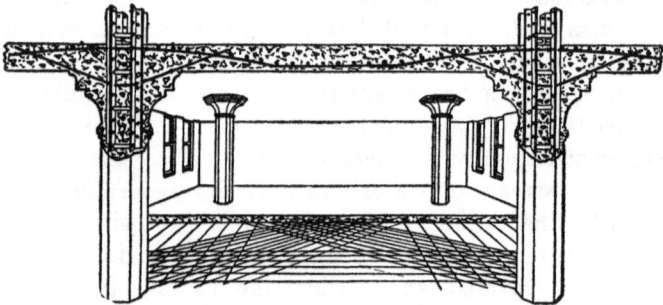

Fig. 66.   Spider-Web System.

tem, including the special advantages claimed for it, are as follows:

This construction has no beams or girders, thus permitting the light to be diffused, unobstructed, over the entire ceiling, and, reflected, to penetrate the interiors of wide buildings;

Permits sprinkler-heads to be economically spaced, which in some cases makes a saving of twenty-five per cent on the cost of equipment;

Permits trolley beams to be placed flat against ceiling, instead of being hung on beams and hangers;

Reduces the height of each story by the depth of beams used in beam and girder construction, which aver-

ages sixteen inches, aggregating twelve feet in the height
of a nine-story building, hence the cost of construction
and heating is reduced.

The steel core of the Barton column is formed of
angles held in position by bolts and pipe separators,
which are designed to carry the load placed on them.
Where desired, spiral steel hooping may be placed on
the outside of metal core.

The formation of the metal core permits the concrete
in column to be rammed at all stages of the concrete
pouring, as the core is unobstructed before the pouring
of the concrete.

Metal centering plates are placed at the tops of
column cores at floor levels, making it impossible for
centers of columns to be otherwise than directly over
one another, forming the steel in continuous columns from
foundation to roof. In the case of rod columns, it is
extremely difficult to place the rods over one another
or make them continuous. This causes eccentric loading
of columns, which accounts for many instances of
collapse.

In the Spider-Web floor and ceiling construction, all
rods are small—usually three-eights inch to one-half inch
in diameter, concealed in a continuous slab, and therefore
fireproof. The octagonal hoop, which acts as a key to the
eight trusses, is eight feet across. In buildings where
columns are spaced twenty feet to centers the distance
between column heads is twelve feet. The steel trusses
comprising column head are usually formed of three-
quarter-inch round rods, and all pass through holes in
the vertical steel forming core of the column. As compared with ordinary beam construction, not only is the
span, in effect, reduced, but the carrying capacity is
further increased, because the spider-web pulls from all
angles when loaded, whereas the beam construction pulls
from only two directions.

Another system of the same general type of

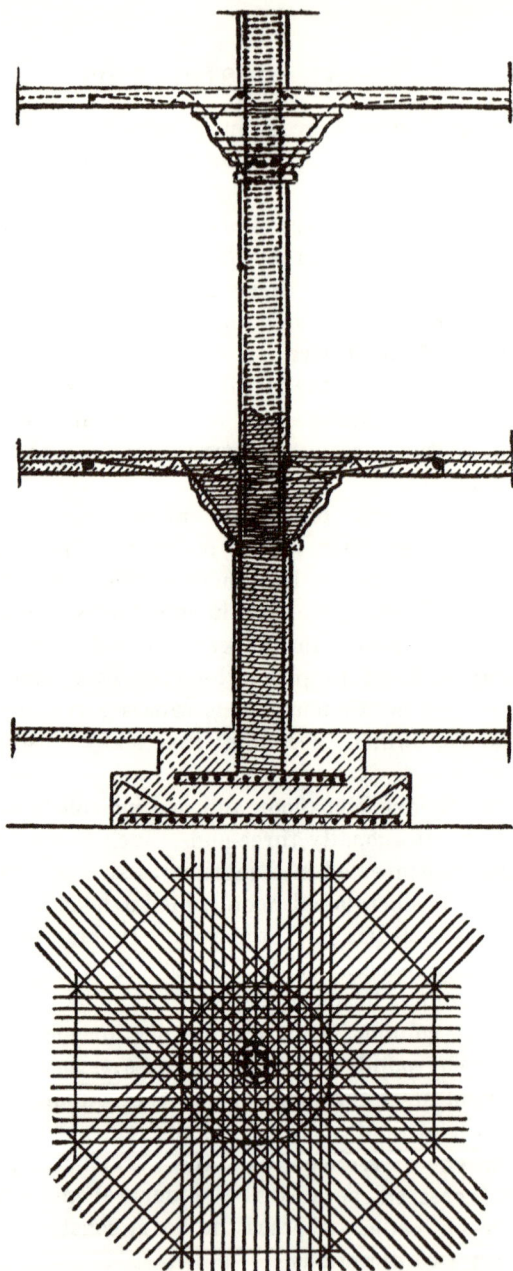

Fig. 66A. Details of Cowles Umbrella System.
See Fig. 66B.

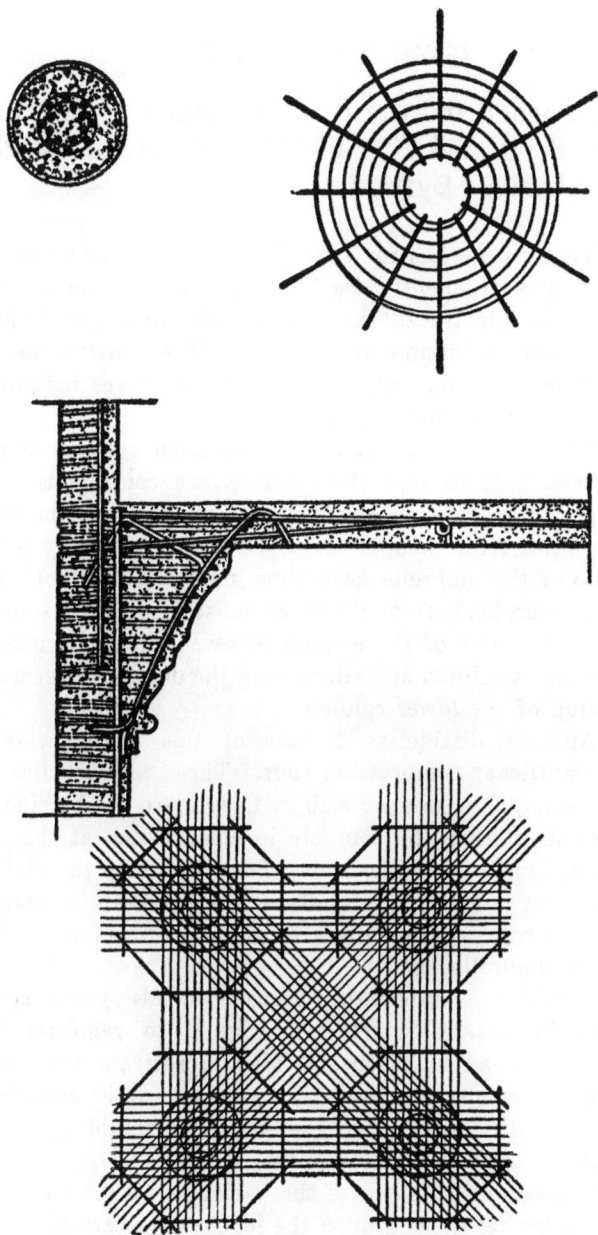

Fig. 66B. Details of Cowles Umbrella System.
See Fig 66A.

the Turner and the Barton systems is shown in Fig. 66 **A** and Fig. 66 **B**. This is the **Cowles "Umbrella" System.**

This system has for its object the doing away of the tee beams frequently used in reinforced concrete construction, and the substitution therefor of flat slabs integral with the supporting column. The construction, it is claimed, is not only economical, but gives maximum strength with minimum material.

This system is a departure from other systems of this general class in that the columns are continuous, and, if spliced, a **telescope splice** is used in the umbrella head, which has triple hooping reinforcement. Not only is the value of the umbrella head thus increased threefold, but the column loads from above are substantially transmitted into the center of the column below, thereby insuring a continuous column and minimizing the danger of eccentric loading of the lower column.

Another distinctive feature of this construction is the **cantilever compression rods.** These rods lie directly beneath the tension or slab rod, and are so distributed as to strengthen the concrete in compression at the perimeter of the umbrella head. They also help the slab at this point in shear, and at the same time help to restrain the concrete in the bottom of the slab forming the top of the umbrella head.

A still more interesting feature of this system is the **umbrella basket,** which is designed to reinforce the umbrella head in shear and also to restrain the entire concrete forming the head. This basket can be assembled and spirally wound, at the shop, preferably upon a machine built for the purpose.

The size and amount of the various steel sections used in this basket depend upon the loads to be carried. This statement, of course, applies also to all of the steel forming the structure,

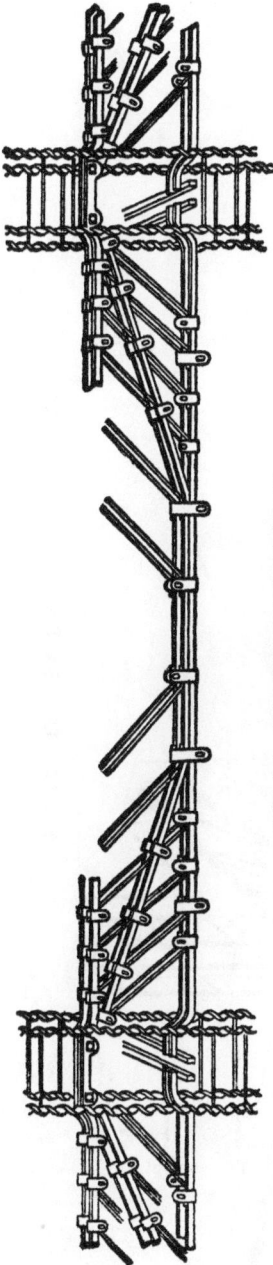

Fig. 67. Pin-Connected Girder Frame.

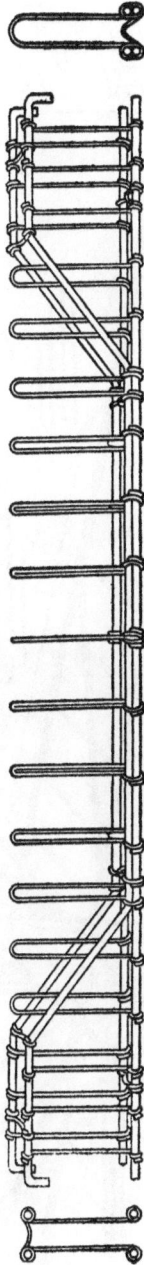

Fig. 68. Unit-Girder Frame System.

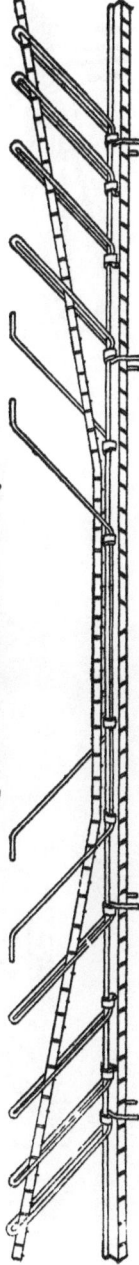

Fig. 69. Herringbone Trussed Bar Used as a Girder Frame.

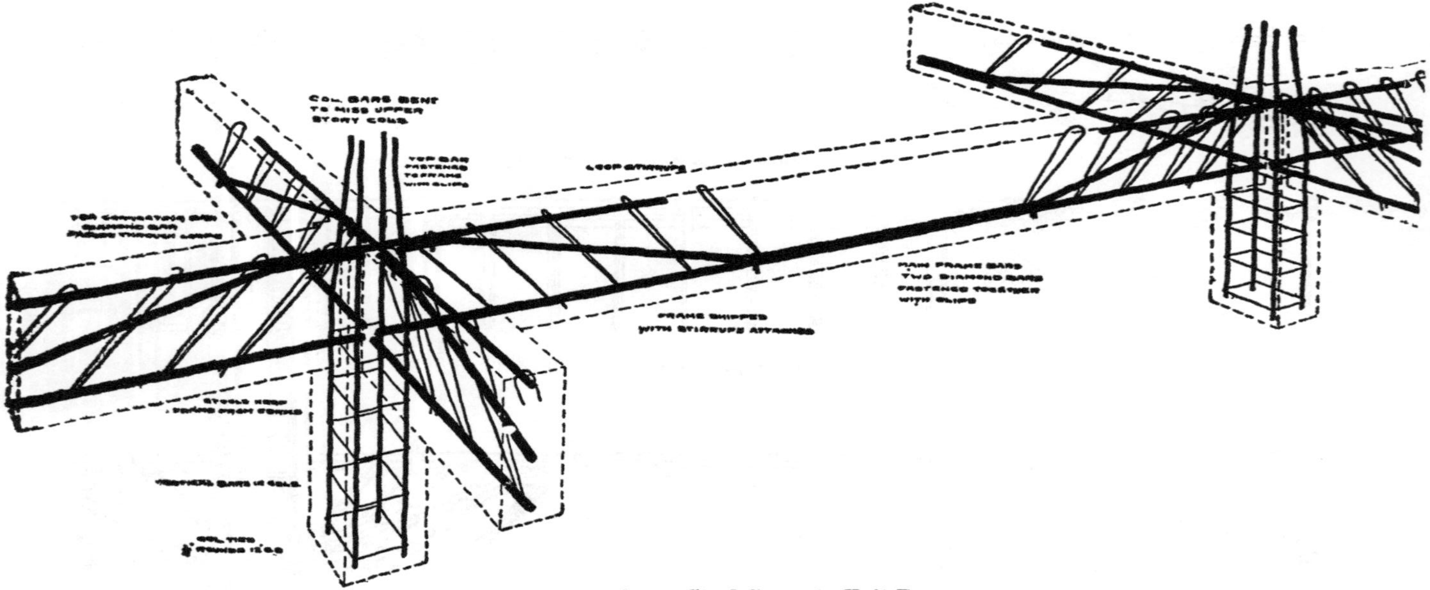

Fig. 70. Lock-Bar Steel Concrete Unit-Frame.

As an example of the **Unit-Beam System** of reinforcement, the **Pin-Connected Girder Frame,** as shown in Fig. 67, is given. These frames are shipped from the manufacturers ready to be lifted into the forms. No expensive field work, either blacksmithing or assembling, is required. Each frame is so numbered that the steel is erected rapidly and economically; and the dangers of misplaced steel, which always attend the use of loose rod reinforcement, are entirely avoided. The mechanical features pointed out as of greatest importance are:

Diagonals rigidly attached at both ends, which may be spaced as frequently as is necessary to resist the shearing stresses;

Carrying one of the main members to the top at the supports, and returning it to provide for negative moments;

A link and pin connection over each point of support, giving each frame a mechanical connection with adjoining frames, so that bonding action of the concrete is not depended upon to transmit stresses from beam to beam.

In use as beam reinforcement, the required amount of steel is made up by using as many frames (units) as are necessary. In the construction of these units bars are used, varying by sixteenths, from $\frac{5}{8}$ of an inch to $1\frac{1}{4}$ inches.

Another specimen of this same type of reinforcement is the **Unit Girder Frame,** shown in Fig. 68.

Fig. 69 shows a **Herringbone Trussed Bar** used as a unit-frame.

Fig. 70 shows a type of unit girder and column frame—the **Lock-Bar Steel Concrete Unit-Frame.**

The essential feature of this system is the shop-made girder and beam frame. A special feature is the loop stirrup, which is tied mechanically and by adhesion, at both the top and bottom, and which is accurately and rigidly fastened in the shop.

The cut shows a beam between columns, reinforced by a single frame. The two bottom bars with stirrups attached are shipped in a unit as shown, ready to set in place. Two adjacent frames being in place held up from the forms by stools of proper height already attached, the top bar is passed through the loops of the stirrups and clamped above the bent bars. The reinforcement is then in place accurately and rigidly.

The bars used are diamond-shaped, with ribs for a mechanical bond, and are set one above the other on their points, and in the direction of the short diameter of the diamond. This serves to reduce the height of the reinforcement, and the ribs serve to keep the bars apart so that the concrete is certain to reach every face of the bar.

By the use of various sizes of bars in the frames, varying from ½ in. to 1¼ in., and by the use of any number of frames in a beam, the reinforcement can be regulated and accurately designed and laid out to meet varying conditions.

For column and slab rods, a round-edge and concave-sided bar is used, which may be plain or twisted as required.

Fig. 71 shows the **System M** form of reinforced girder and beam construction. This system consists of an independent light steel frame, which is subsequently so united to concrete that the two become a reinforced construction.

The upright iron or steel supports and the steel floor members can be formed with sections of any shape suitable to make a light, independent structure capable of maintaining its form. The steel uprights are erected plumb and the steel floor members are set level. The girders are connected to the uprights to keep them plumb, and the floor-beams are connected to the girders to make a sufficiently rigid structure.

This steel frame forms the guides for the erection of the wooden or other moulds into which the concrete forming the floors, and forming the covering of the uprights, is to be poured. The reinforcement cannot be displaced or misplaced by careless workmen.

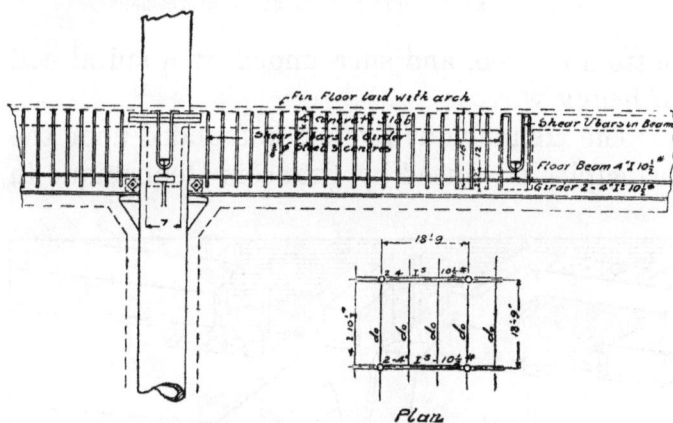

Fig. 71.   System M.

Although any kind of steel section which is sufficiently stiff can be used for floor members, and several kinds of special sections offer best theoretical results, still good, practical results have been obtained by the use of small I-beams and channels. The size of the I's and channels used, it is claimed, is usually not more than one-half the height of the steel beams required, for ordinary construction, to carry the entire floor load. In this system the smaller the height of the steel members (in concrete beams of the same depth), the greater the economy of steel, on account of the increased effective depth of the reinforced concrete beam.

The **De Vallière System** for beam and girder construction is shown in Fig. 72. This system consists of a main reinforcing bar located near

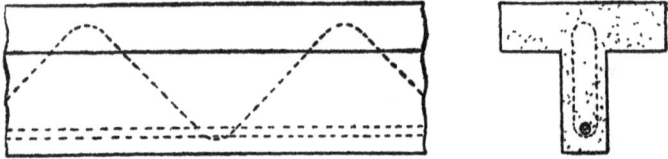

Fig. 72. De Vallière System of Reinforcement.

bottom of web, and surrounded by a spiral coil of heavy wire.

The **Hennebique System** is founded upon the principles advanced by François Hennebique in

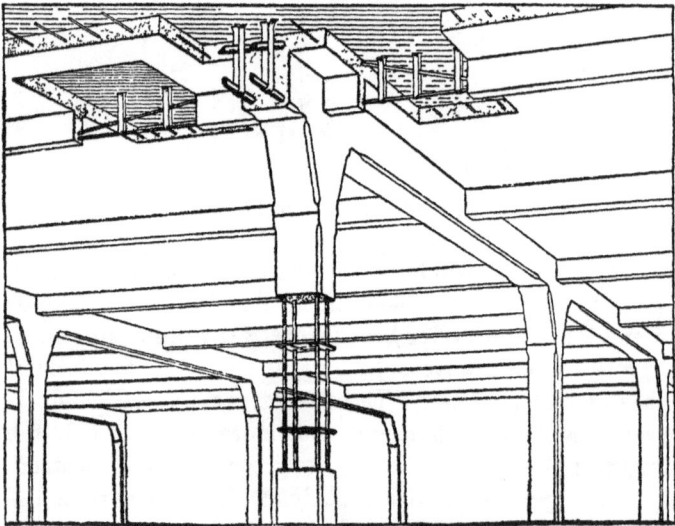

Fig. 73. Hennebique System.

the early stages of reinforced concrete work. Fig. 73 shows the general plan of arrangement of reinforcing. Other views of this system are given later under "General Building Construction."

In the Hennebique beam the concrete is relied upon to resist the compressive stresses in the upper part of the beam, while the steel rods resist all tensile stresses in the lower portion. The concrete also forms the connection between the two flanges, assisted by the stirrups, generally formed of hoop steel. These are of great impor-

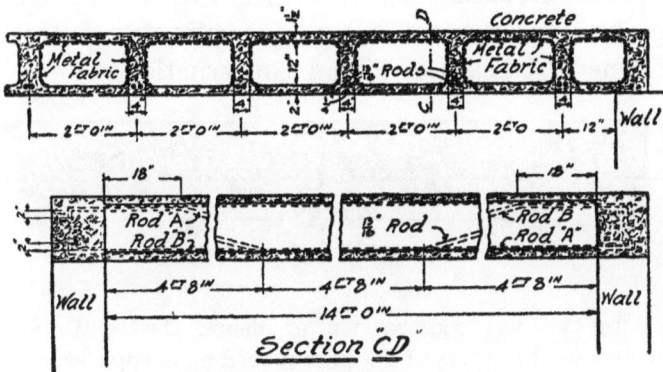

Fig. 74. Merrick Floor System.

tance. Besides acting as a connection between the upper and lower parts of the beam, they take also the horizontal shear. For this reason the stirrups are placed closer together near the supports.

The rods in a beam are of two kinds—straight and bent. The bent rods, besides taking their proportions of the direct tension and of the shear, take any tensile stresses in the upper portion due to a negative moment over the support. A beam of this description is similar in many respects to a timber trussed with steel tie-rods and brackets.

The **Merrick System** for reinforcing floors is shown in Fig. 74. The object of this system is to lighten the weight of the concrete slab. Directly upon the forms a 2-inch layer of concrete is placed; and, before this has set, oblong boxes of metal fabric of small mesh are laid horizontally, with the reinforcing rods in the spaces between them, and the concrete is filled in between the boxes and around the reinforcing rods, and covered over the top to form the floor.

The **Roebling System** (see Fig. 75) is employed in connection with a structural steel frame of I-beam or girder construction.

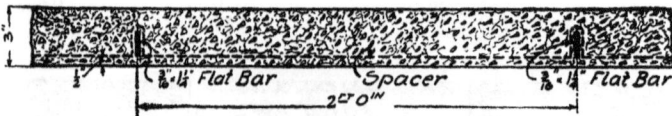

Fig. 75. Roebling Floor System.

For all flat construction of floors, the reinforcing system used consists of flat bars placed upon edge, secured at the ends to the steel beams and bridged with bar separators. The object of the edgewise position of the bars is the increased protection thus secured to the reinforcing steel. With this type of floor the structural steel frame is generally completely encased with concrete.

For light roof construction where the steel work need not be protected, a continuous slab is built over the beams, reinforced with flat steel bars, 3/16 in. by 1¼ in., placed edgewise and held in position by spacers, as shown in Fig. 75.

For floor construction this system also uses segmental arches of cinder concrete laid upon permanent stiffened wire lath centering, or upon wood centering which is car-

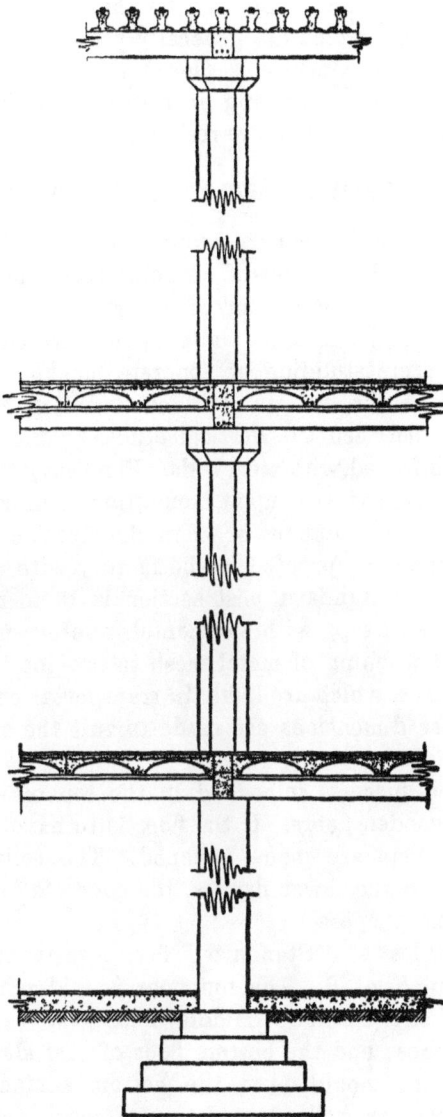

Fig. 76. Three Styles of ''Standard'' Floor Construction.

ried on steel tees and supported by the steel I-beams of the floor system, which are generally placed about 7 feet on centers. The concrete is placed upon the centering without puddling or tamping, in order to obtain a light porous concrete of high fire-resisting quality.

The **Standard System** is shown in Fig. 76.

The general scheme is to build floors of light-weight I-shaped or T-shaped joists of reinforced concrete to replace wood joists or reinforced concrete slabs, and rest the ends of the joists upon walls made of vertical interlocking concrete studding or concrete blocks. Columns are formed in the wall in light construction by filling the hollows between the vertical studs, or blocks, with concrete reinforced with steel rods. For heavy buildings, the floor-joists may rest upon monolithic reinforced concrete girders and columns, or upon structural steel girders and columns fireproofed in the factory with concrete. One T form of standard joist section is 16 inches wide by 8½ inches deep, with horizontal reinforcement for tension, and webbing of metal mesh to provide for shear and the stresses which are liable in transportation. Members of other dimensions are made to suit the span and loading required.

A nailing piece is imbedded in the top of the joist for laying wooden floors. If the floor is to have concrete finish, the joists are made I-shaped. The ceilings are plastered upon the lower flanges, the concrete being left rough for the purpose.

Three styles of "Standard" floor construction are illustrated in Fig. 76. The top floor is laid with joists just described; the two middle floors, of separately moulded arches; and the bottom floor of cast slabs, with reinforced ribs moulded on the bottom surface. The thin slabs are also well adapted to roof construction.

An important feature of the "Standard" system is the method of connecting the individual members. The

PLACING CONCRETE FLOOR IN IRRIGATION CANAL, UNCOMPAHGRE PROJECT, COLORADO.

View during construction—Forms in place.

View after forms were removed and before dressing up.

Courtesy of Office of Public Roads, Washington, D. C.

CONSTRUCTION OF CONCRETE BEAM BRIDGE AT CHERAW, S. C.

reinforcement is allowed to project, and is mechanically connected after placing. The connection is finally im bedded in fresh concrete so as to give strength and rigidity.

In Europe the **Siegwart System** of floor construction has been developed quite extensively, using for floor slabs a series of adjacent hollow beams formed by the use of collapsible cores.

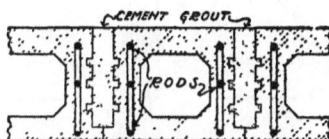

Fig. 77.   Siegwart Floor System.

Fig. 77 shows the construction of these hollow beams, and also their reinforcing rods.   The top face of these beams forms the floor slab, and the bottom face forms the ceiling for the rooms below.   These beams are moulded individually in sections about ten inches wide.   As shown by the figure, they have an open space extending nearly to the lower surface on each side of the beam, or, are narrower across the top than across the bottom.   When they are placed in position in the structure, these open spaces are filled with cement grout, thereby making the floor practically one continuous mass.

The **American System** consists in the use of plain round rods underneath woven wire fabric, both of **high-carbon steel**. Fig. 57 shows the type of fabric, and the clamp used in joining

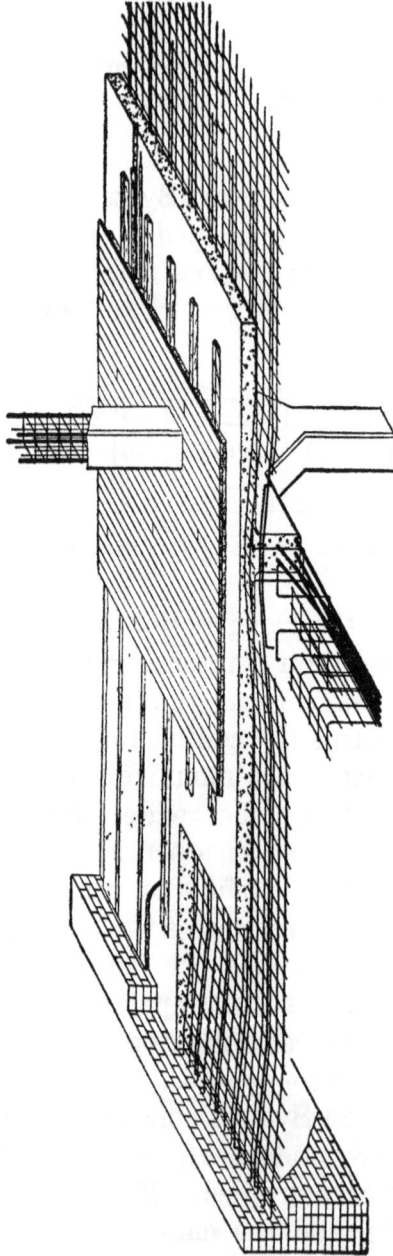

Fig. 78. "American" System of Reinforcement.
Showing practical application of rods and fabric to girders, columns, and floor and roof slabs.

ends of sheets. Floors are designed in some instances by stretching the parallel rods across steel girders, and simply laying the wire fabric on top of these, embedding all in concrete. In other structures, the monolithic type of caged concrete column and connected girder, all formed of rods and fabric embedded in concrete, is used, as in Fig. 78. In short spans and for light loads, the wire fabric is often used without the rods.

The manufacturers of these materials claim advantages on account of ease and cheapness of handling and laying the rolls of fabric, together with the absolute assurance of proper spacing of reinforcing and ideal distribution of metal.

The **Vaughan System** of floor construction is shown in Fig. 79. Other details of this system are shown later under the heading of "General Building Construction."

The Vaughan system consists of reinforced concrete "joists" made in an exaggerated I-beam shape. For ordinary construction, the upper and lower flanges are 12 inches wide, and the total height of joist is 8 inches. The upper flange is made somewhat thicker than the lower. Any good system of reinforcement combining tension and horizontal shear members may be used.

The joists are laid in the building with the flanges touching, forming a continuous and complete surface on top to hold the finishing floor. Any type of floor—plain cement finish, tile. granolithic, or wood—can readily be laid on the joists.

The under side of the joists forms another complete and continuous surface for the ceiling, requiring but a single coat of rough plaster on which to place the final finish.

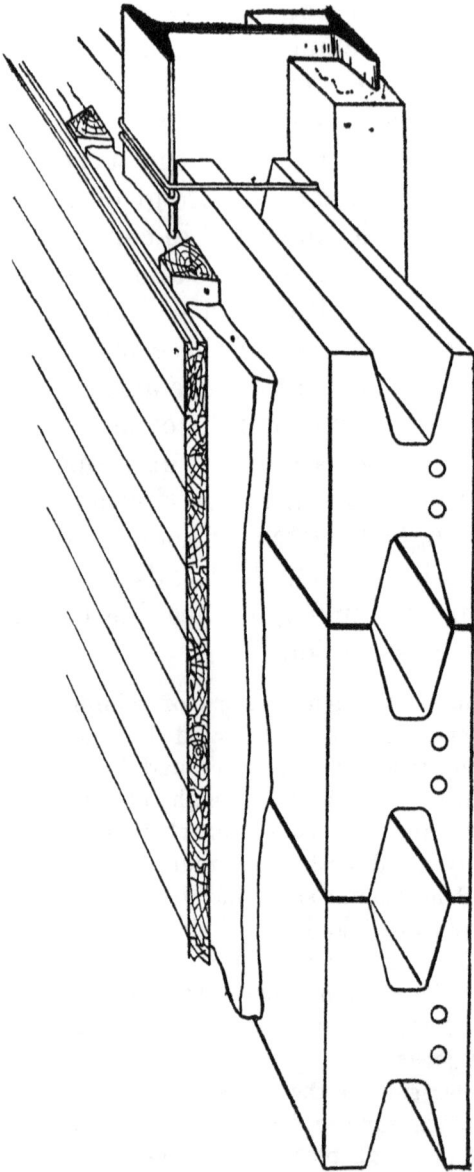

Fig. 79. Vaughan System.

The hollow space between the joists forms a convenient space for the laying of pipes, wires, etc., as well
as providing a large air space which prevents dampness,
sound-proofs the floor, and acts as a non-conductor for
heat and cold.

The use of this system does not limit in any way the
other structural features of the building. Any wall can
be used except the ordinary wood studding, which, of
course, would not be suitable for fireproof floor construction. The use of two-piece concrete blocks, using them
only on the inside as a bearing wall for the joists, makes
a specially good and cheap wall. The outside may be
veneered with brick, concrete blocks, or tile, or finished
with metal lath and stucco as may be desired.

For intermediate supports a simple construction is the
use of steel I-beams and cast-iron columns.

The joists are cast on the ground, either at a plant
established for the purpose, or may be made at the point
where they are to be used. Made in forms of wood or
pressed steel, readily admitting variations in length, and,
if necessary, changes in depth and proportions to meet
different conditions of loading, they present a form of
concrete construction adaptable to greatly varying conditions.

After curing from 20 to 30 days, the joists are ready
to be placed in the building, although, as in all forms of
concrete construction, they will keep on increasing in
strength for a much longer period.

This method of construction does away with all falsework, staging, or centering, the use of hollow tile, etc.,
for filling. A still further saving is due to the absence
of all obstructions in the interior of the building, and in
consequence the opportunity to push the inside finish
rapidly.

Fig. 79 shows the Vaughan system joists
used with an I-beam for the intermediate sup-

port. The cement shoe tile shown are a stock product. They are cast in two sections, and form a fireproofing coat for the beams as well as a support.

In factories and warehouses, a layer of concrete one inch in thickness on top of the joists is all that is necessary to finish the floors. When a wood floor is required, it can be readily placed as illustrated. A grillage of light wood strips is laid directly on the joists, and a 1½-inch layer of wet concrete poured in between to hold them in place. This leaves a space directly under the wood floor for the passage of pipes, electric wires, etc.

The **National System** for the reinforcement of concrete makes use of a special device for locking the rods or bars together.

Fig. 80 shows the rods or bars in place, fastened together with the **National Chair,** on the form. No part of the rods or bars, either cross or longitudinal, rests upon the form, thus permitting them to be thoroughly imbedded in the concrete.

Fig. 81 illustrates the ''chair'' for locking the heavier wires and bars, the lower right diagram showing it assembled with the wedges turned up, making a perfect lock. The wedges are turned up with a very simple tool. The cut also shows the feet of the chair which keep it from sinking into the form. In spiral or column construction the feet of the chair are absent

The above systems are a few of the many in use for reinforcing the **superstructure** of a building—that part above the foundations. The methods of reinforcing **foundations and footings**

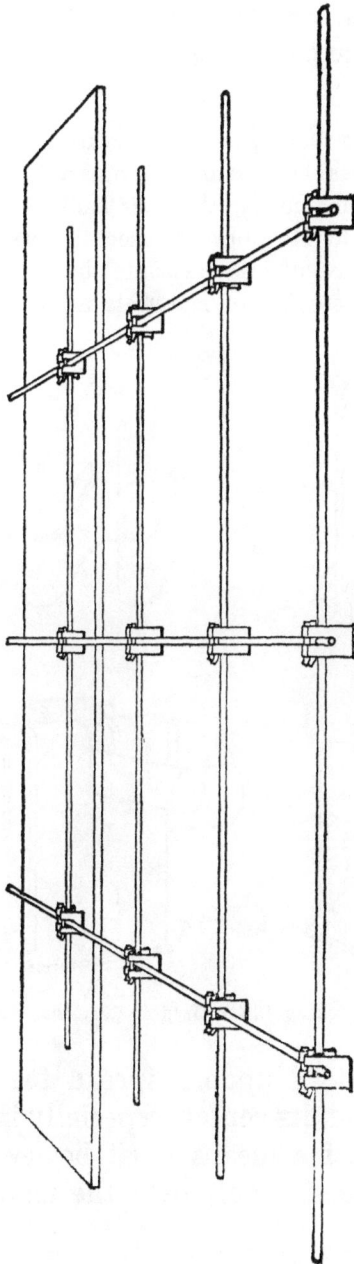

Fig. 80. The National System.

for buildings resting upon soft or unstable ground may be considered as coming under two classes:

1. When the footing is to be spread out wide enough so that the allowable pressure per square foot of bearing area is a safe amount for that particular soil;

2. When concrete piles are used to obtain the same result with the use of less horizontal bearing area and a corresponding lessening of materials used.

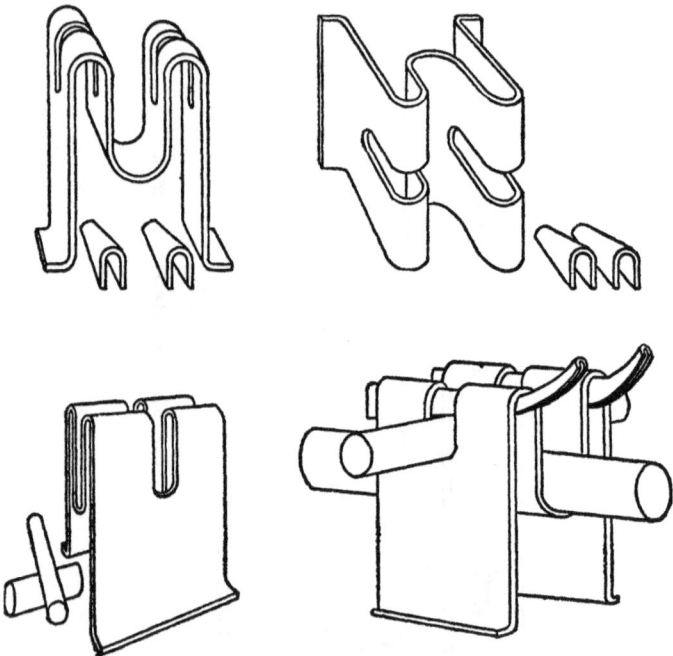

Fig. 81. Locking Chair Used in National System.

Since the load upon a spread footing commonly comes on its center, especially in the case of columns, some means must be used to distribute that load evenly over the under side of

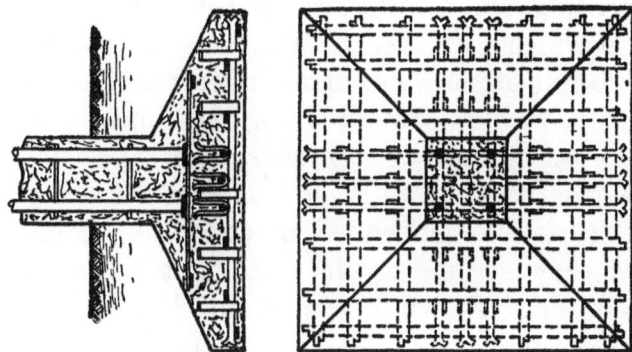

Fig. 84. Hennebique Column Footing.

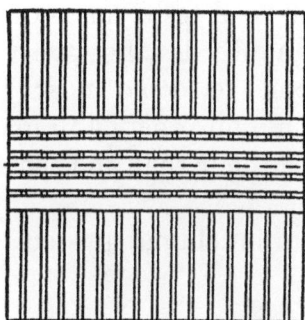

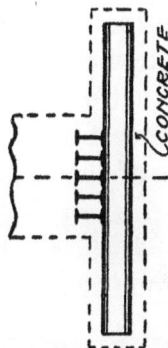

Fig. 83. Reinforced Concrete Footing of I-Beams.

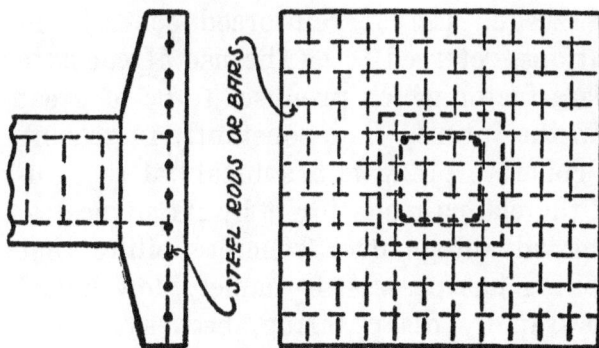

CONCRETE

STEEL RODS OR BARS

Fig. 82. Reinforced Concrete Column Footing.

the footing, and also to provide for any bending action which may occur in the wide flat surface on account of the concentrated central load. A common way of providing for these necessary points is by the use of rods, bars, I-beams, T-beams, old steel rails, etc., laid in a form of grill work and embedded in concrete. Sometimes they are simply laid at right angles to each other as shown in Fig. 82 and Fig. 83; and at other times diagonal layers are used, in addition to the former layers.

These footings do not need to be made of an equal thickness throughout, as the greatest bending action will tend to come near the middle; therefore it is common practice to make the edges of the footing thinner than the central part.

Fig. 84 shows the Hennebique system of column footings.

CONCRETE PILE FOUNDATIONS

The second use of reinforced concrete in foundations, referred to as the use of **concrete piles,** is a factor which promises to be of great value in the future. The constantly increasing use of concrete piling is a substantial recognition of the claims made for it by its advocates. The one advantage over concrete piling that wood piling has possessed—namely, low initial cost—is rapidly disappearing, because of the growing scarcity of the available lumber supply due to constant deforestation. The absolute permanence of concrete piling, its freedom from the dangers that threaten the integrity of wood piling—rot, over-driving, the attacks of boring animals, etc.,—its low ultimate cost, and the fact that its constituent materials may be obtained almost anywhere, are factors that in time will drive wood piling out of general practice.

There are two general methods of concrete pile construction: those **constructed in place,** and those **moulded or rolled in advance and driven** by methods similar to those used for driving wooden piles. In the first method some form of collapsible steel core, encased in a closely fitting shell of suitable material, is driven in the usual manner, the core withdrawn and the shell filled with concrete. Piles constructed in this manner may be either plain or reinforced, depending largely upon the character of the work for which they are to be used. In the second method it is quite essential that they be reinforced to withstand the strains to which they are subjected in handling and driving.

In certain cases concrete piles are an economical substitute for deep pier foundations. Six types of patented reinforced concrete piles are shown in the accompanying illustrations.

The **Simplex pile** is constructed by driving a hollow shell with a point to the full depth, and gradually raising the shell as the concrete is placed in the hole thus made. The process, using an **alligator point,** which opens when the shell is pulled, is shown in Fig. 85 at the left.

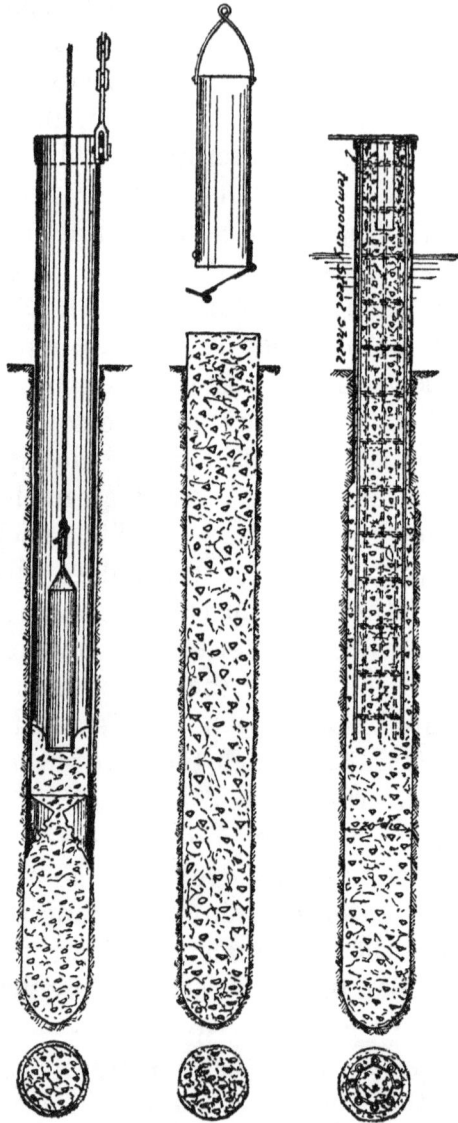

Fig. 85. Simplex Piles.

Sometimes a solid point made of cast iron is used, which is left in the ground.

Raymond concrete piles are made by driving a tapering sheet steel shell to refusal by means of a collapsible steel core, withdrawing the core, and thereupon filling the shell with concrete.

The shell consists of a number of circular sections that are formed by uniting the vertical edges of two pieces of 18 to 20 gauge sheet steel, bent into shape by a cornice brake. The diameters of the sections range in a decreasing ratio from the uppermost section down to the point or boot. The latter is stamped from a single piece of 16 gauge stock.

The core is composed of three steel segments forming a tapering cylinder or cone. The segments are separated or brought together through the action of a series of wedges. A driving cap is attached to the head of the core.

The shell is assembled by slipping the various sections composing it over the core, the segments of which are expanded at this stage. Placing the boot in position over the point of the core completes the shell. The sections overlap sufficiently to exclude any soil, water, or other foreign substances that might otherwise gain admission into the shell while it is being placed.

After the core is completely encased in the shell, it is driven to refusal. The core is thereupon withdrawn by bringing the segments together, or "collapsing" the core, as the operation is termed. The shell, which is of sufficient strength to retain its shape after the withdrawal of the core, remains permanently in the ground and forms a mould or form for the concrete. Fig. 86 shows the core of a Raymond pile collapsed and partly withdrawn from the shell.

Before being filled the shell is subjected to careful inspection. After inspection, thoroughly mixed concrete,

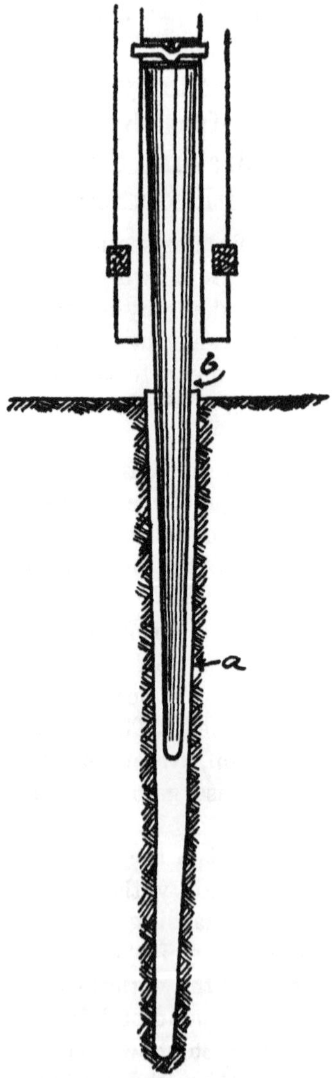

Fig. 86. Raymond Pile.
Showing driven shell.

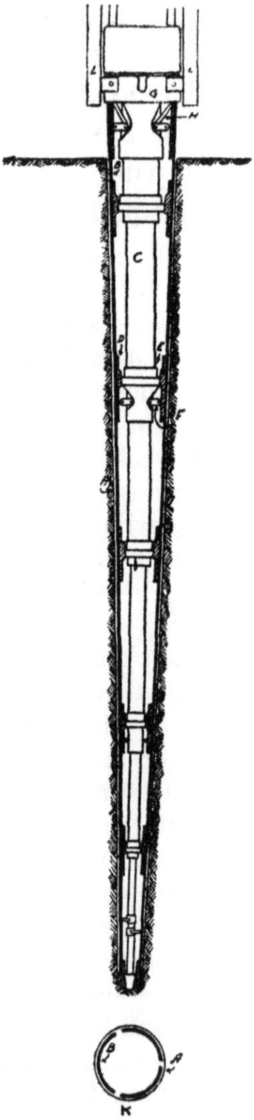

Fig. 87. Section of Raymond
Pile Core.

composed of one part good Portland cement, three parts
sharp sand, and five parts crushed stone or gravel of
suitable size, is poured in, being carefully tamped until
the shell is filled.

Fig. 87 shows a section of the Raymond pile core.
A represents the shell, driven in the ground; B, the
exterior plates (¾ in. thick) of the pile core; C, the stem
of the pile core, made of extra heavy pipe of diminishing
diameter as the lower end of the pile core is approached;
D, the wedge-shaped castings fitted to the exterior plates;
E, the corresponding wedges fitted to the interior stem
(this wedge is made of a steel casting, which also acts as
a collar for coupling together the various sizes of pipes
forming the stem); F, hinges linking the exterior plates
and interior stem of the core (note their position when
core is expanded); G, the head of the core, made of cast
steel hollowed out at the top to receive an oak cap block,
which receives the blow of the hammer; H, keys to keep
the exterior plates in place when expanded; K, cross-
section, showing opening between plates to allow for collapsing; L, steel leads of driver.

Fig. 87 is a sectional view of the core, showing collapsing and expanding device. (Steam hammer in the
leads resting upon the core.) In this illustration the shell
is driven and the core expanded.

The reinforcing of concrete piles with steel rods
is sometimes found to be desirable. In the case of the
Raymond pile, the insertion of the reinforcing material
is done prior to the placing of the concrete. This operation is extremely simple, is done in plain sight, and
requires no unusual skill.

Should a concrete bond be desired between the butts
or heads of the piles and the superstructure, that portion
of the pile projecting above the surface of the ground is
stripped of the surrounding shell. Thus, a larger bonding
surface is offered than could otherwise be obtained.

The Gow pile, as shown in Fig. 88, has an en-

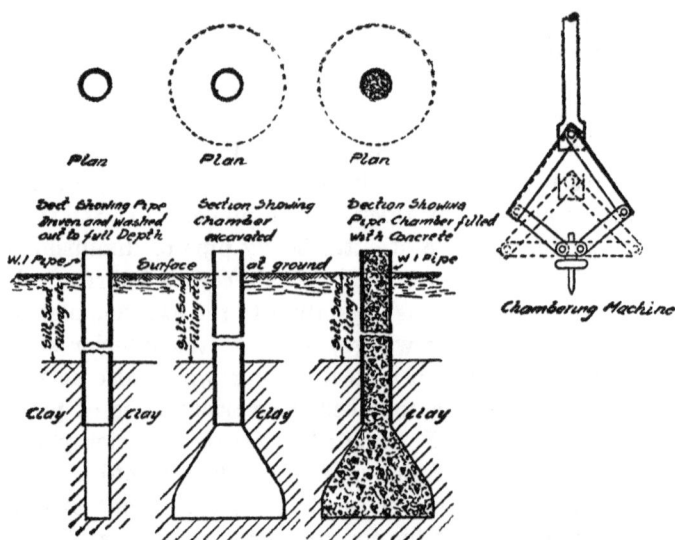

Plan   *Plan*   *Plan*

Sect Showing Pipe   *Section Showing*   *Section Showing*
Driven and Washed   *Chamber*   *Pipe Chamber filled*
out to full Depth   *excavated*   *with Concrete*

W.I Pipe   *Surface*   *at ground*   *W I Pipe*

Clay   *Clay*   *Clay*   *Clay*

Chambering Machine

Fig. 88.   Gow Method of Concrete Piling.

larged footing so as to give it larger bearing,
and is formed by washing down a tube with a
water jet to a firm strata, and then enlarging
the bottom of the excavation by an expanding
arrangement to form the base of the pile. The
apparatus is withdrawn, and the space filled with
concrete.

In many cases where too many boulders are
not liable to be encountered, piles of rectangular or round shape are built horizontally upon
the ground, reinforced with steel rods, and, after
setting for at least a month, are driven with a
pile-driver. A special form of cap is required
to break the force of the ram on the head of the

A CONCRETE HIGHWAY BRIDGE AT CHELMSFORD, MASS.

A CONCRETE ROAD CULVERT AT AMHERST, MASS.

RESIDENCE BUILT OF STUCCO ON METAL LATH.

GARAGE BUILT OF SOLID REINFORCED CONCRETE.

pile. The corrugated pile is a special type of driven pile.

Fig. 89 shows a typical cross-section of the **Gilbreth corrugated pile.** This particular arrangement of the corrugations was adopted for reasons of economy in the construction of the forms. At first glance it looks like a circle, but in reality it is made as an octagon or a hexagon, with grooves that do not diminish at the small end of the pile, although the pile tapers from

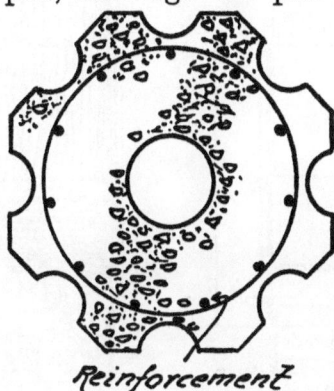

Fig. 89. Cross-Section of Gilbreth Corrugated Pile.

16 inches across at the larger end to 11 inches at the smaller, lower end. The reinforcement shown is electrically welded fabric, and the size is approximately 3/8-inch wires, 3 inches on centers longitudinally, with approximately 1/8-inch wires, 12 inches on centers around the pile. The hole in the center is made 3½ inches in diameter at the top of the pile, and 2 inches at the bottom of the pile. This hole is made tapering for two reasons:

(a)   So as to have a large quantity of concrete in the lower end of the pile.

(b)   So that the tapering plug that is used to cast the hole in the pile can be easily withdrawn.

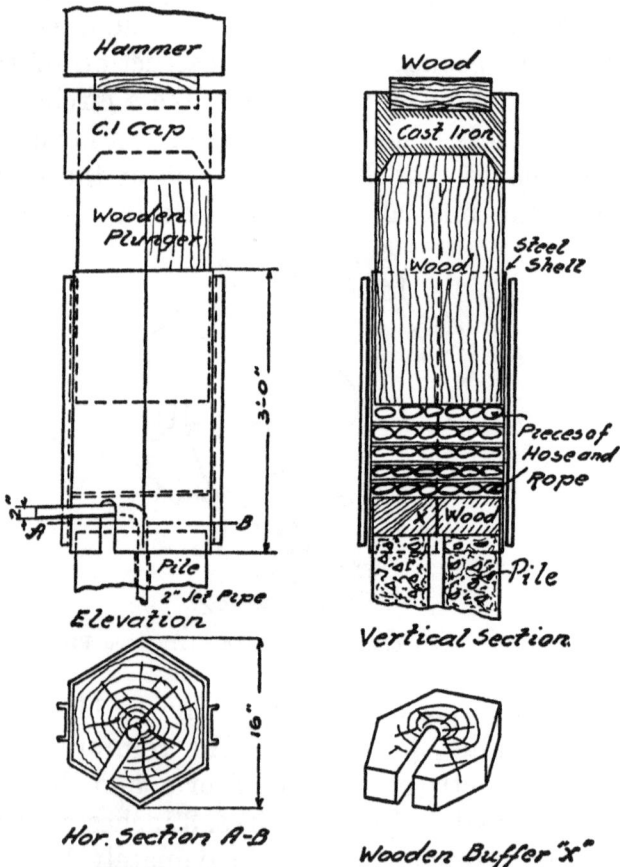

Fig. 90.   Details of Cushion Cap for Driving Gilbreth Corrugated Piles.

This pile is driven by means of a jet of water and a heavy weight, the weight to be used as a hammer when necessary. The pile-drivers used are of the ordinary type used for wooden piles. The piles are hoisted into place ready for driving; the cushion head, as shown in Fig. 90, is lowered over it; the water jet placed down into the center of the pile; and the pressure turned on. The jet extends through the entire length of the pile and protrudes three inches below the bottom of the pile. The tremendous pressure of the water is sufficient to dig a hole and carry the loosened sand, gravel and earth up the corrugations, which act as an exhaust to the jet. The weight of the hammer pushes the pile down into the hole. When the pile is nearly in place, the hammer and cushion cap are hoisted up, and the jet is removed, the cushion cap is again lowered over the head of the pile, and the hammer forces the pile to refusal. A large amount of alternate layers of old rope and rubber-lined canvas hose forms a cushion sufficient to protect the pile from injury in driving.

In the moulding of these corrugated piles, one side of the hexagonal or octagonal mould is left off, so that the concrete can be properly tamped and thoroughly inspected at all times by the foreman and the inspector. One of the features of this pile is that there is no guesswork regarding its construction, or regarding the location of its reinforcement.

In the **Chenoweth System** of concrete piles and cross-ties for steam and electric railway work, no forms are used. A reinforced concrete pile is produced simply by rolling a sheet of concrete and metal netting into a solid cylinder, as shown in Fig. 91. Piles thus formed can be made in long lengths and a wide range of diameters. The principle involved is that the compressive value of concrete is many times greater than its tensile strength. In this construction, the reinforcing members act independently, the concrete carrying the load action under a compressive stress, while the steel reinforcement resists the tensile stress.

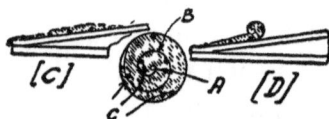

Fig. 91. Detail of Chenoweth Pile.

Referring to Fig. 91, **A** is a steel pipe or rod; **B** is a coiled sheet of wire netting; and **C** are longitudinal steel rods, placed near the periphery, equal distances apart, parallel to the longitudinal axis of the pile. The pipe or rod **A** forms the shaft or mandrel on which the pile is rolled or wound, and the netting **B** and rods **C** constitute the reinforcement.

The apparatus for rolling the pile consists of a traveling platform, between which and a roller the pile is formed. The winding pipe or mandrel is set in line of the shaft of a large spur wheel. In operation the steel wire netting with the longitudinal rods attached is spread on the platform and covered with a layer of concrete, as indicated by the sketch **C**, Fig. 91. One edge of the netting is attached to the edge of the platform, and the other edge to the winding pipe or mandrel. Thus arranged, the winding mandrel is rotated, and the netting and its covering of concrete are wound or coiled up as indicated in sketch **D**. At the same time and as fast as

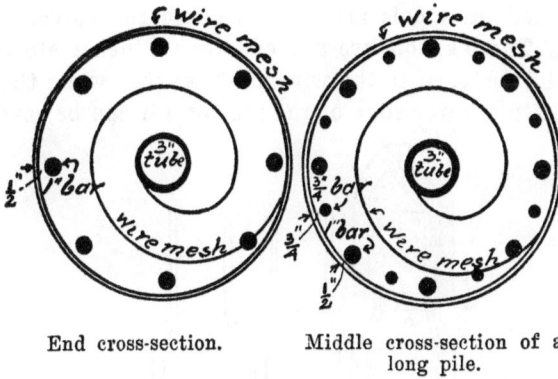

End cross-section.   Middle cross-section of a
long pile.

Fig. 92.   Diagram Showing Reinforcing Members of Chenoweth
Concrete Piles.

the netting and concrete are coiled up, the platform **A**
moves under the roll **B** and the roll itself rotates.

The forming cylinder of concrete is by this means kept

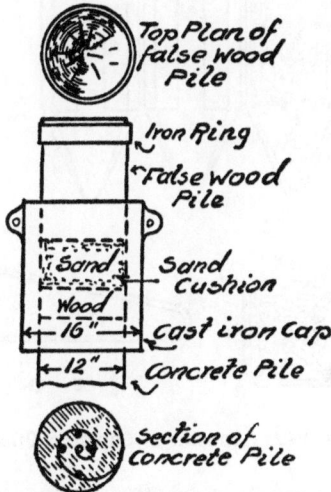

Fig. 93.   Cap Used in Driving Chenoweth Pile.

under constant and heavy pressure between the platform
and roller. To bind the roll of concrete, wires are wound
around it at close intervals and tied. The wires are

contained on spools arranged beneath the moving platform. These spools are spaced every 6 inches along the
pile and fastened to the wire mesh, so that when the pile
reaches the lower edge of the platform it can be revolved

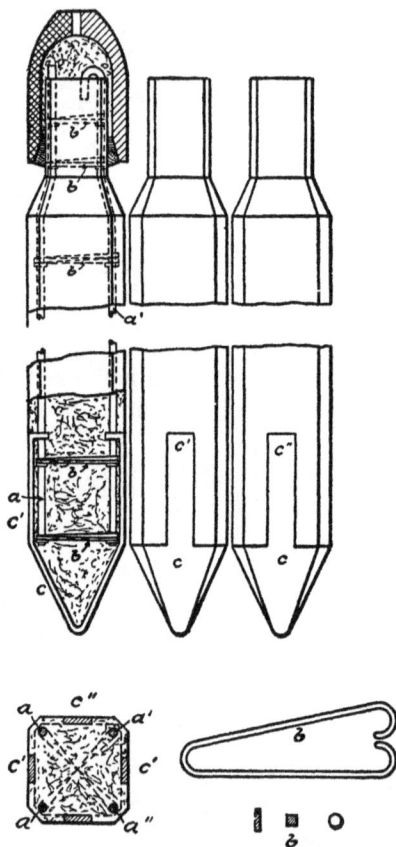

Fig. 94. Hennebique Concrete Piles.

about the axis, causing the wire to be wound on the pile.
In this manner the wire is passed around the pile a
number of times and secured. At the driving end of the
pile, the spools are placed 4 inches apart for the first
6 feet of the pile, and then 6 inches apart for the

remainder of its length. The pile is removed from the machine by rolling onto a car, when it is transported to a platform, where the point is formed, flushed up, or smoothed, if desired, and allowed to remain until it hardens.

Fig. 92 shows the reinforcing members of a Chenoweth concrete pile 14 inches in diameter and all lengths up to 75 feet.

Fig. 93 shows the detail of cap used in driving these piles. It was found that this cap would protect even the sharp edges of the top of the pile, provided the pile was made eight days before driving.

Hennebique piles (Fig. 94) are similar in construction to the Hennebique columns. The point, however, is protected by a cast-iron or sheet-metal shoe. In the sheet piles the point is formed by beveling one of the narrow faces so as to form a wedge with the opposite face, and both narrow faces are slotted by a semicircular groove running from point to butt. The semicircular grooves form, when the piles are driven as a sheeting, a circular hole from top to bottom between each pair of piles. This hole serves for a water jet during driving; and when finally filled with cement, the whole forms a monolithic sheet of piling. The Hennebique piles are driven by a drop-hammer in the same manner as wooden piles; the top of the pile is, however, protracted by a false pile or cap, which serves to lessen and distribute the shock.

General Building Construction

The primary consideration in any design is that the finished structure shall serve, in the most adequate manner, the purpose for which it is built. This will determine to a large extent the location of the columns, the general framing and type of the floor construction, depth of girders, size of columns, etc. The design of a warehouse built to carry heavy loadings will vary, accordingly, from that of a residence. In the former, girders and beams placed close together might be used to advantage; in the latter, the appearance of unsightly beams in the ceiling of a room may be considered faulty design. Similarly in a factory the layout of the girders may be planned so as to accord with shafting for machinery.

A bridge is similar in this respect, as its purpose, location, and amount of waterway will predetermine to a large extent its design. The engineer or architect will find, however, that reinforced concrete, owing to its plasticity, lends itself admirably to every possible requirement and condition.

A manufacturer about to build a factory or warehouse must choose between several types of construction. In this selection the governing considerations are **cost, safety, durability, and fire protection,** while many minor factors enter into each individual case.

Types of buildings for mills, factories, and warehouses may be classified as follows:

(1) Frame construction;
(2) Steel construction;
(3) Mill or Slow Burning construction;
(4) Reinforced concrete construction.

The first and cheapest type—**frame construction**—may be neglected as unsuitable for permanent installation, because of its lack of durability and its fire risk. Board walls, narrow floor-joists, board floors and roofs, not only do not protect against fire, but in themselves afford fuel even when the contents of a factory are not combustible.

Steel construction with concrete or tile floors, provided the steel is itself protected from fire by concrete or tile, is efficient and durable; but its first cost alone will usually prohibit its use for the ordinary factory building.

Mill construction—or **slow burning** construction, as it is sometimes called to distinguish it from fireproof construction—consists of brick, stone, or concrete walls, with wooden columns, timber floor-beams, and thick plank floors, which, although not fireproof, are all so heavy as to retard the progress of a fire and thus afford a measure of protection.

Reinforced concrete, through the reduction in price of first-class Portland cement and the greater perfection of the principles of design, has lately become a formidable competitor to both steel and slow burning construction—a competitor of steel, not only for factories and warehouses, but also for office buildings, hotels and apartment houses, because of its lower cost, shorter time of construction, and freedom from vibration; a competitor of slow burning construction because of its greater fire protection, lower insurance rates, durability, freedom from repairs and renewals, and even, in many cases, its lower actual cost.   Another condition favorable to the increase of concrete buildings is the increasing scarcity of yellow pine from the Southern States.   The price of all lumber has increased so much in recent years that the cost of a first-class mill-constructed building is almost as high as the cost of a concrete structure.   Comparative bids recently taken in some instances show only a difference of 5 per cent greater cost for the concrete over mill construction, and in other cases 10 per cent; so it is probable that the average difference in cost for a first-class mercantile building is somewhere between 5 and 10 per cent.

Considered from a hygienic point of view, reinforced concrete offers the advantage of **absence of porosity,** which is inseparable from the present ordinary modes of building.   It does not afford shelter to rodents (rats, mice, etc.); they cannot attack reinforced concrete.   This property is greatly appreciated in flour mills, grain warehouses, etc.

Reinforced concrete can also be **easily and thoroughly cleaned**—a property of great value in hospitals, schools, barracks, etc. Jail cells and banking vaults have been constructed from reinforced concrete, and even bullet-proof screens for target practice. It may be noted here that several great powers are now using reinforced concrete for the protection of fortresses against shell.

ERECTION OF REINFORCED STRUCTURES

The erection of a reinforced structure is similar to the making of a casting in a large foundry. Forms or patterns are built to correspond exactly with the lines of the finished work, the reinforcing steel is set in place, and the concrete is poured into the forms. The whole structure is thus built as a monolith, and moulded into the finished form. The concrete is allowed to set a requisite length of time, the forms are removed, and the building stands complete—a structure carved, as it were, out of solid rock.

It is thus seen that the erection work consists primarily of four distinct operations:

(1)   The erection of the centering or falsework;
(2)   The placing of the reinforcing steel;
(3)   The mixing and placing of the concrete;
(4)   The removal of the centering.

As the **forms** represent the mould from which the finished structure is made, great care is used to make these exact and true to line. They are built rigid, and thoroughly braced so as to bear

the weight of the plastic concrete without deflection. In order to give a smooth finished surface, planed boards are used, and corners of columns and beam boxes are chamfered. All joints
are set closely together to make the forms fairly
water tight. Soft soap or some like preparation
should be smeared on the forms to prevent the
concrete from sticking.

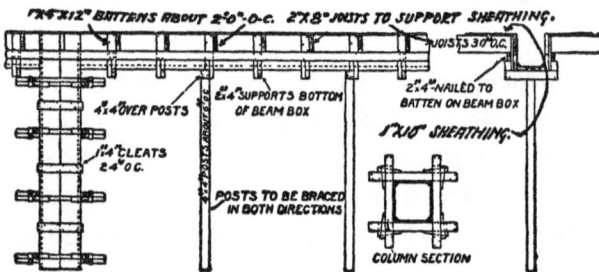

Fig. 95. Beam and Column Forms.

The steel is set accurately in place in accord-
ance with detailed drawings prepared for that
purpose; and these drawings are followed
explicitly.

Only the best materials should be used for
the concrete, and these are thoroughly mixed in
the proper proportions. A rather wet mixture
is used. The concrete is poured into the forms
and is laid continuously over the entire floor
area. It is placed carefully around the steel
work, so as not to disturb the location of the bars
and to cover them thoroughly at all points. The
concrete is puddled in the form so as to allow no
voids to occur,

TABLE XIII

Approximate Mixtures Adaptable to Various Classes of Work
Rich, 1·2:4; Medium, 1: 2.5:5; Ordinary, 1:3:6; Lean, 1:4:8

(A. S. Johnson.)

KIND OF WORK	MIXTURE	CONSISTENCY
Abutments	Rich to Ordinary	Medium
Arches	Rich to Medium	Medium
Backing for Masonry	Lean	Medium to Dry
Beams, Reinforced	Rich to Medium	Very Wet
" Plain	Rich to Medium	Very Wet to Medium
Cisterns	Rich to Medium	Very Wet to Medium
Columns, Reinforced	Rich	Very Wet
Conduits, Water	Rich	Very Wet
Coping	Rich to Medium	Medium
Culverts, Reinforced	Medium to Ordinary	Medium
" Plain	Medium to Ordinary	Medium
Driveways	Same as Sidewalks	
Fence Posts	Rich	Very Wet to Medium
Floors, Reinforced	Rich to Ordinary	Very Wet to Medium
" Ordinary Ground	Medium to Ordinary	Medium
Footings	Ordinary to Lean	Medium
Foundations, Heavy Vibrating Machinery	Rich	Very Wet to Medium
" Ordinary Machinery	Medium	Medium
" Thin Walls	Rich to Medium	Very Wet to Medium
" Thick Walls	Medium to Lean	Medium to Dry
Girders, Reinforced	Rich to Medium	Very Wet
" Plain	Same as Beams	
Gutters	Same as Sidewalks	
Pavements	Same as Sidewalks	
Piers	Rich to Ordinary	Medium
Reservoirs	Rich to Medium	Very Wet to Medium
Roof Slabs	Medium to Ordinary	Medium
Sewers, Reinforced	Rich to Medium	Medium
" Plain	Medium	Medium
Sidewalks (Base)	Medium to Ordinary	Medium to Dry
" (Sub-Base)	Ordinary to Lean	Medium to Dry
Silos	Rich to Medium	Very Wet to Medium
Tanks	Rich to Medium	Very Wet to Medium
Walls, Dwelling Houses	Rich to Medium	Very Wet to Medium
" Large Buildings (Compression and Tension)	Rich to Medium	Very Wet to Medium
" Large Buildings (Compression Only)	Medium to Ordinary	Medium
" Massive	Medium to Ordinary	Medium
" Retaining	Medium to Ordinary	Medium
" Thin Foundations	Rich to Medium	Very Wet to Medium
" Tunnel	Medium to Ordinary	Medium

The hardening of concrete is not a "drying-out" process, as some suppose, but is a chemical action caused by the addition of the water to the cement. The concrete takes its "initial set" in a short time, and therefore should be deposited in place as quickly after mixing as possible.

Concrete work is often carried on in the winter months, and will freeze if precautions are not taken. The freezing retards the setting of the concrete, and often completely ruins it. It is usually best to remove any concrete known to have been frozen. Simple precautions can be taken to prevent such freezing—such as heating the materials, adding salt to the water (less than a 10 per cent solution), keeping the building heated by charcoal grates, and covering the concrete after being laid with some good insulating material such as cement bags, straw, manure, etc.

After the concrete has thoroughly set and hardened, the forms are carefully taken down. This is done gradually and evenly so as to cause no undue shocks to the concrete work. The length of time necessary to leave the forms in place depends very largely on the atmospheric conditions, the season of the year, the thickness of the concrete work, and the kind of cement used. With the removal of the forms the structural portion of the building is complete and ready for use. The concrete, however, will continue to grow harder and stronger every day.

Table XIII gives the proper mixtures and degrees of wetness for various forms of work.

FLOORS, SLABS, AND ROOFS

When some one of the materials for reinforcement already described is placed in a construction in parallel rows, spaced equally and covered with concrete, the result is a **reinforced slab.** If these rows run only one way, the construction is called an **independent bar reinforcement.** If they cross transversely, the result is a **latticed reinforcement.** Latticed reinforcement is well adapted to floor and roof work as the transverse members of the reinforcement prevent shrinkage cracks.

Examples of independent bar construction are shown in the plain I-beam reinforcement illustrated in Fig. 4 (next volume); also in many of the bar constructions, such as the Ransome in Fig. 61 or the Kahn in Plate 14.

The latticed construction is shown in the use of plain rods laid transversely in a slab, as in Fig. 82; in the use of wire fabrics consisting of rods connected transversely by lighter rods or wires, as in Fig. 58; or in the use of expanded metal as in Fig. 53. The crossed rods in the floor slabs of the Turner, Barton, and Cowles systems might also come under this heading.

Reinforced concrete floor construction may again be divided into three classes:

(1) Those constructions which serve simply as a filling between the girders and beams of a floor framework of steel;

(2) Those in which the girders and beams are themselves reinforced concrete;

(3) Those in which the girders and beams are done away with.

The first class may be divided again into **flat slab** and **arched slab** constructions. As examples

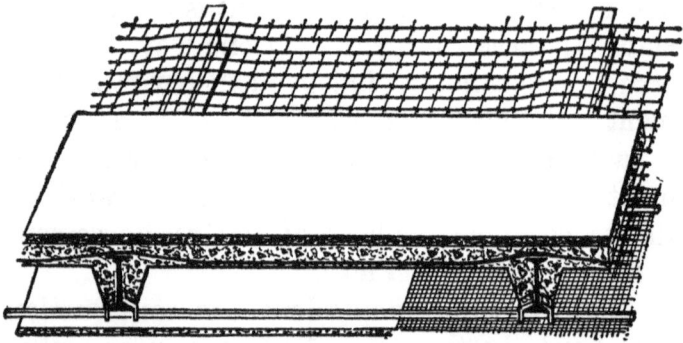

Fig. 96.   Sheet Fabric as Used in Concrete Slabs.

of the flat slab floor construction, we have the
Roebling system (Fig. 75), the **Merrick system**
(Fig. 74), the **Vaughan system** (Fig. 79), the
National system (Fig. 80), the **Kahn system**
(Fig. 96), and numerous other systems based on
the use of expanded metal or woven wire fabrics.

As examples of the arched slab construction,
see Fig. 97 and Fig. 98.   Both of these figures
illustrate the use of a closely meshed reinforcing
agent which in many cases proves of great as-
sistance in constructive work.   With a small
mesh material, the concrete may be dumped directly in upon the previously bent up forms, just
enough of the concrete passing through the
meshes to provide a rough surface for plastering

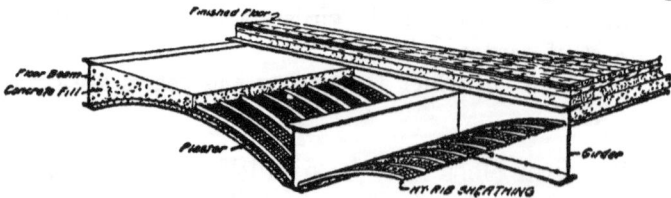

Fig. 97.   Arched Floors.

SPRING CANYON FLUME, INTERSTATE CANAL, NORTH PLATTE
IRRIGATION PROJECT, NEBRASKA.

Built of reinforced concrete.

WASTEWAY GATES AND PORTAL OF TUNNEL, HUNTLEY IRRI-
GATION PROJECT, MONTANA.

GATES AND SLUICEWAY, YUMA IRRIGATION PROJECT, LAGUNA, CALIFORNIA.

WASTEWAY AT MOLLIES FORK, INTERSTATE CANAL, NORTH PLATTE IRRIGATION PROJECT, NEBRASKA.

Nature's work in solid rock is here supplemented by the art of the modern worker in concrete.

on the ceiling formed by the under side of the arch. Ends of the bent-up sheets of reinforcing rest upon the flanges of the I-beams shown, or are fastened to the beam reinforcing rods in case of the construction shown in the lower part of Fig. 98. No centering or forms for the arch are necessary in this construction.

In case a reinforcing agent of such a nature were used that it would not hold the wet concrete mixture in place, forms would have to be employed as in regular floor construction, and the reinforcement placed in them in the usual

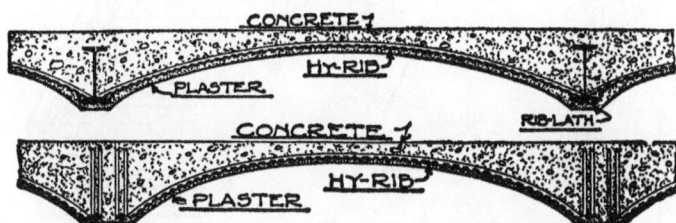

Fig. 98. Arched Slab Floor Construction.

manner. The arched type of floor is used for carrying heavy loads.

In these structures of the first class, the floor-slab either rests on top of the beam or girder, embeds the top flange, or rests upon the bottom flange.

The second class comprises floors of such a construction that the girders and beams really constitute ribs for strengthening the slabs. These are monolithic constructions. As examples of this type, we have the **Hennebique system** shown in Figs. 99 and 99**A**; the **Gabriel**

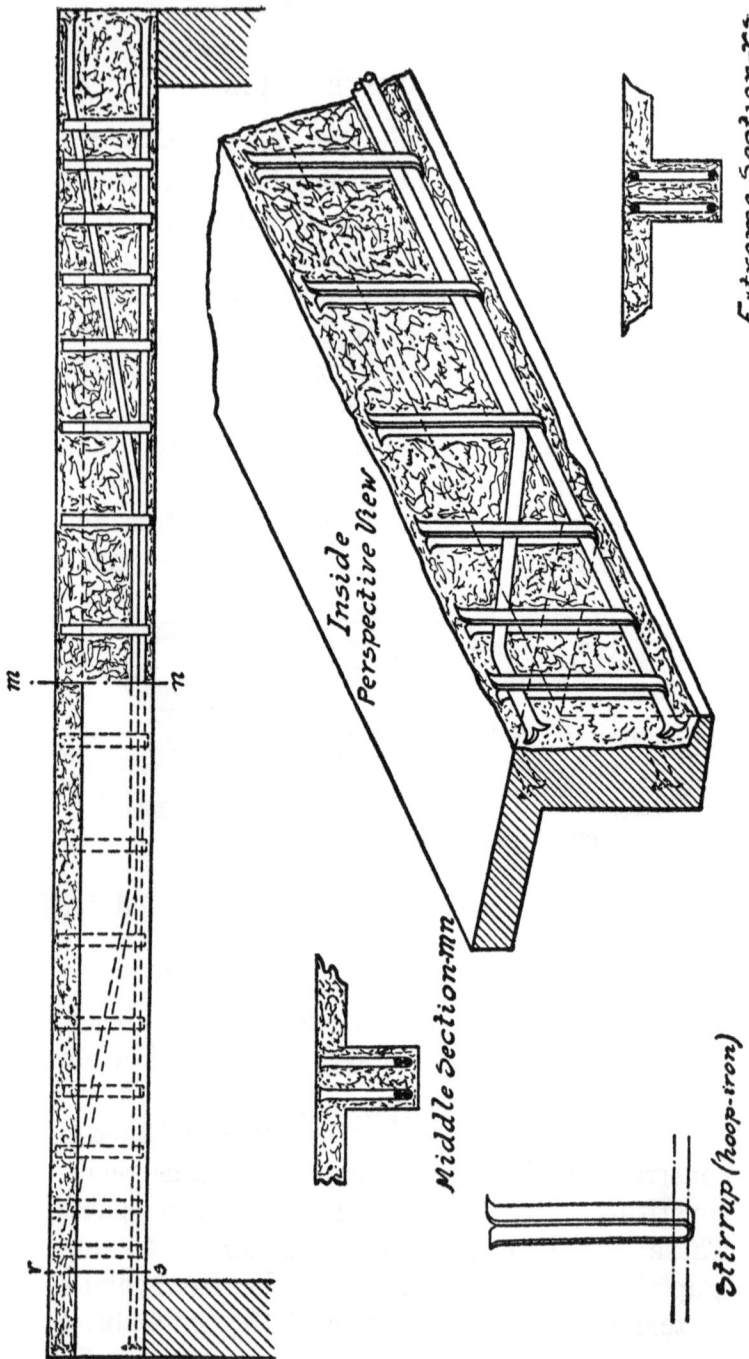

Inside
Perspective View

Middle Section-mm

Stirrup (hoop-iron)

Extreme Section-rs

Fig. 99. Typical Hennebique Beam.

m

n

r

s

system shown in upper part of Fig. 100; the **Kahn system** shown in Plate 15; the **American system** shown in Fig. 78; the **Unit-girder systems,** and many others shown in the previous pages.

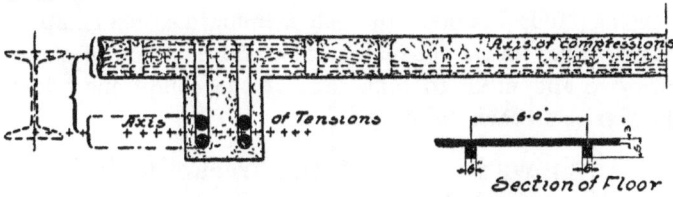

Fig. 99A.   Hennebique Flat Floor Construction.

The third class, comprising those monolithic structures in which the beams and girders do not project, are illustrated by the **Turner Mushroom system** shown in Plate 16; the **Barton Spiderweb system** shown in Plate 17 (general floor

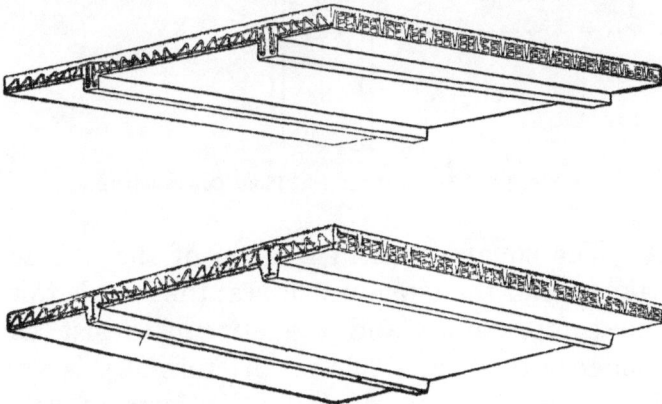

Fig. 100.   Reinforced Concrete and Hollow Tile Floor Construction.

layout); the **Cowles Umbrella system** shown in

Figs. 66**A** and 66**B**; and the **Smith Girderless Floor Construction** shown in Fig. 101.

A brief description of the **Smith system** is as follows:

A unit-column consisting of spiral wire column reinforcing; four brackets to each column, also made up of spiral wire; and a continuous floor fabric made of a size of wire and mesh to take care of the conditions which may arise on each building.

As shown in Fig. 101, the bracket is designed to take care of all stresses which may come upon

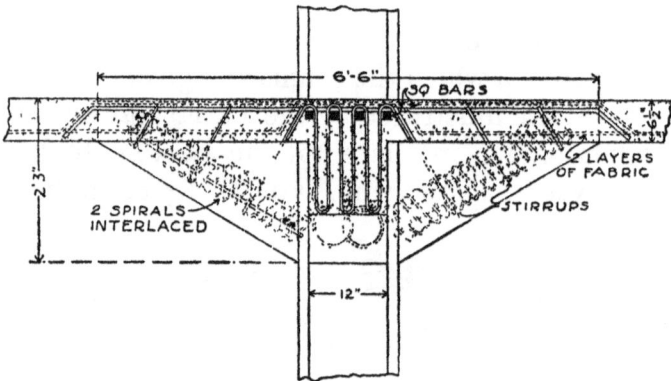

Fig. 101. Smith Girderless Floor Construction.

it. The square bars in the top of the bracket take care of the tension stresses; the spirals take the compression; and the stirrups resist the tendency to shear. The proprietors of this system claim to have devised this form of reinforcing to eliminate the cost of assembling the steel in the field, and furthermore, to have done away with the danger of unskilled workmen

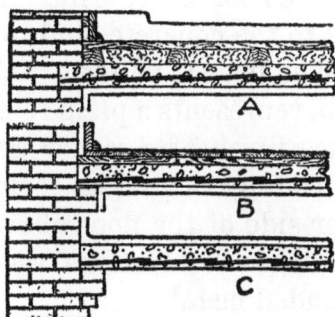

Fig. 102. Three Forms of Floor Construction.

placing the steel in the forms in a haphazard
manner—a very common occurrence and the
cause of many failures. The columns of this system are a unit for each story; the brackets are
assembled in the shops ready to be placed in the
forms, and the fabric for the floor is made in
rolls the full length or width of the building.

The three cross-sections, Figs. 102, **A**, **B**, and
C, illustrate different methods of supporting reinforced concrete floors on brick walls, and
different methods of finishing the floors. In
A and **B**, the plastering is applied directly to the

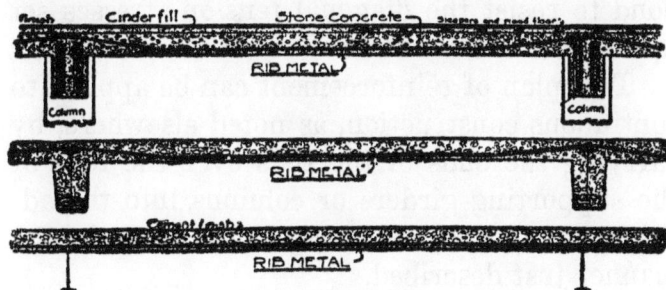

Fig. 103. Typical Short-Span Floor-Slabs.

under side of the floor slabs. In **B** an under-floor is first nailed to the concrete, and to this under-floor the finished floor is nailed.

Fig. 102, **C,** represents a plain reinforced floor with a surface finish of a richer grade of concrete. The ceiling below is formed by the cleaned under side of the floor above.

In each of these illustrations, the reinforcing agent is **expanded metal.**

Fig. 103 shows three types of short-span floor-slabs reinforced with **rib metal.**

Fig. 43 shows an **anchored end bar reinforcement** which is claimed by some engineers to be a suitable construction in beams, girders, and thick slabs. The bars in this form of reinforcement are round, with threaded ends fitted with nuts and large plate washers at each end. The bars are bent to shape before placing in the forms, and rigidly held in place. The number used depends upon the size of the member and the percentage of steel required. The anchor-plates at the ends of the rods prevent serious slipping of the rod in the concrete, and provide a bond to resist the diagonal tension stresses set up in the beam.

This plan of reinforcement can be applied to continuous construction, as noted elsewhere, by carrying the ends of the rods over the tops of the supporting girders or columns into the adjacent span, and anchoring them there in the manner just described.

The principles governing the design of **rein-**

forced concrete roofs are similar to those for floors. The reinforcing materials generally used for roofs are of light weight. The general construction is one of a combination of light steel shapes for the framework, upon which is laid some form of sheet fabric such as expanded metal, rib metal, closely woven wire fabrics, etc. The closer woven metals, when used, do away with the use of wooden forms, thus allowing the concrete to be deposited upon the metal directly; the holes in the fabric letting enough of the wet mixture pass through to form a good bond, and also to provide a rough surface to hold the plas-

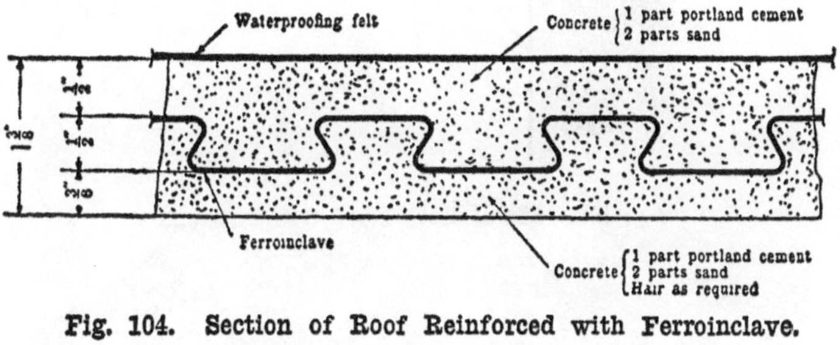

Fig. 104.   Section of Roof Reinforced with Ferroinclave.

ter or finish on the under side. Fig. 104 shows a section of roof made by using a corrugated sheet metal. After the sheet metal is in place and firmly fastened and supported, it is spread over or plastered with cement mortar, forming a sheet or slab 1⅜ inches thick. The figure shows the general proportions of materials used in the mortar, and also the final covering of waterproof felt.

Fig. 105 shows a method of laying ribbed

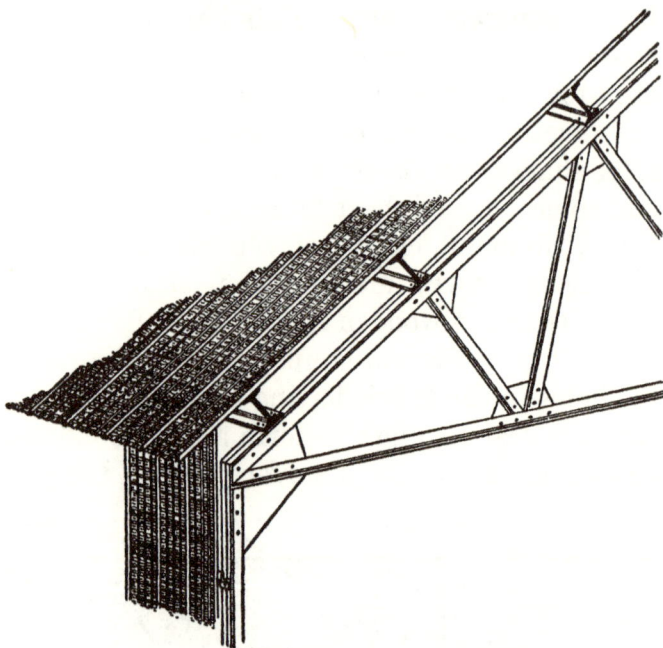

Fig. 105. Metal Lath Roof Construction.

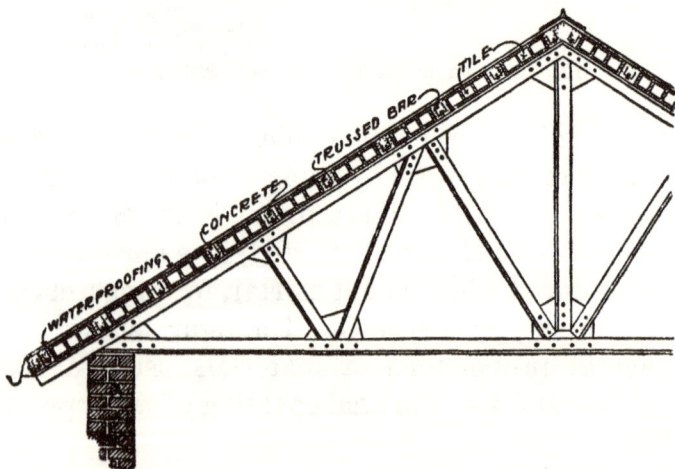

Fig. 106. Tile and Reinforced Concrete Roof.

metal on a steel frame. In the use of ribbed
metal, the rib should be laid upward on the roof.
This places the bulk of the metal on the tension
side of the slab, and also presents a smoother
surface underneath to plaster on. Rib metal or
woven wire fabrics should be fastened firmly to
the structure at frequent intervals—say, every
24 inches, clips being provided for fastening to

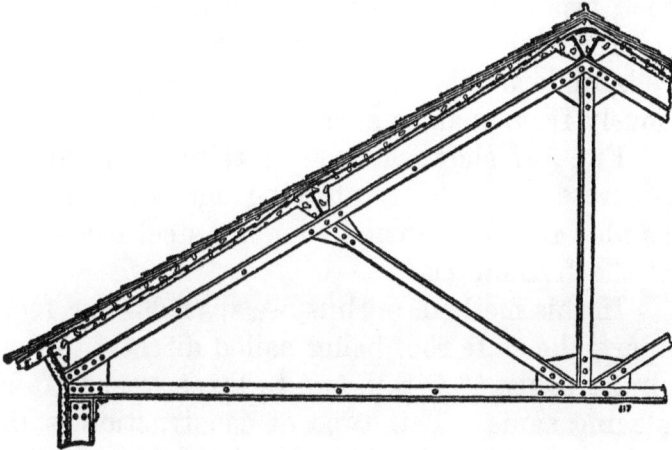

Fig. 107.   Cinder Concrete Roof.

steel members. If a wooden frame is to be covered, the sheets may be nailed directly to the
wood.

Fig. 106 shows a typical form of **tile roof** reinforced by concrete beams. The structure itself
is of steel. The use of tile lightens the weight
of the structure, yet does not interfere seriously
with its strength or insulation against heat or
cold.

When the coarser mesh grades of wire fabrics

and wire-connected rods are used in roofs, forms are built below as in ordinary floor construction, the fabric laid over the supporting steel beams, and the concrete applied as in floor and slab work. An example of this form of construction is shown in the Peoples Gas Light & Coke Company Building in Chicago. Here a span of 10 feet, with roof live load of 50 lbs. per square foot, was filled with 4-inch slabs of concrete reinforced with "American fabric" without any rods underneath. The fabric extended continuously from span to span.

Fig. 107 shows a form of reinforced **cinder concrete roof,** the reinforcing agent being expanded metal stretched over the steel members of the framework.

In this method, purlins are spaced 5 to 8 feet apart, the slate roof being nailed directly to the cinder concrete within two to three weeks after placing same. This type of construction is in extensive use in the Navy Yards of this country and is regarded as being very satisfactory.

REINFORCED CONCRETE COLUMNS

A reference to the figures shown in the pages devoted to the different systems of reinforcement will show the general types of columns used in reinforced concrete work.

The **general plan** of all reinforced columns seems to be that of a cage supported by upright members and filled with concrete, the outside being likewise protected from fire and corrosion by a thick layer of concrete. This cage may be made up in various ways. Two of the common ways are as follows:

(a) By the use of some form of coil of metal surrounding the vertical bars;

(b) By the use of tie-rods spaced regularly up the length of the column.

A few additional details will now be shown, that were not shown in the previous figures.

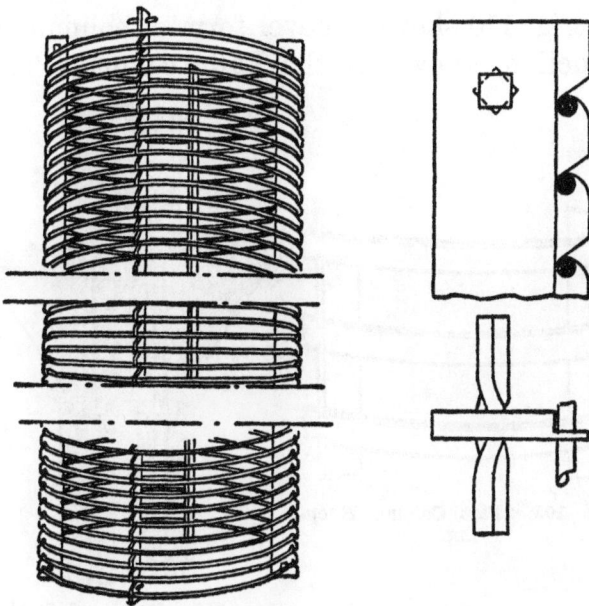

Fig. 108. Details of Smith Spiral Column.

Fig. 108 represents the "Smith" form of column hooping. A special vertical reinforcing bar is used with this hooping, and is shown in detail. The reinforcement consists of a continuous spiral of cold-drawn wire of high elastic limit, rigidly held in place and alignment by clinching into the bar.

Fig. 109 shows the Kahn form of flat bar column hooping. The hooping material is ½ inch by ¼ inch, and can be procured in coils of any diameter. The spacing bars are ¼ inch by 1½ inches, and are punched to receive the coil. The hooping is wedged into place and held rigidly by compressing the spacing bars.

Fig. 110 shows a novel form of column construction, known as the **Lally column.** These

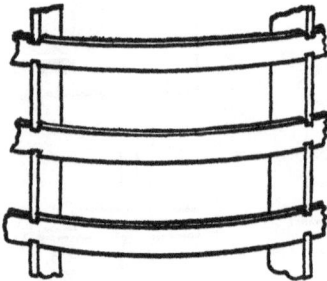

Fig. 109. Flat Column Hooping.

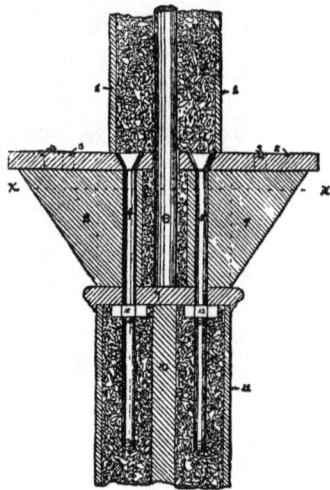

Fig. 110. Detail of Steel Cap for Lally Column.

columns are factory-made by special machinery which, it is claimed, eliminates all air-holes or cavities. The outside shell is of steel; the filling is of sand, cement, and blue trap-rock, with an inner core of steel.   The details shown in the figure are as follows:

1—Represents steel shell of upper column.

2—Represents crown-plate of bracket-cap upon which beams or girders rest.

3—Holes for bolts to fasten beam to plate 2.

4, 5—Represent tie-bolts passing from crown-plate 2, through brackets 8 and 9, also through cap 7, entering the casing 11 of the lower column, and also embedded in the concrete.

6—Represents a steel rod or pintle embedded in the concrete at the base of the upper column, extending into the cap of the lower column, resting on cap-plate No. 7; thus holding the upper column firmly in position.

7—Represents a cap-plate which sets on casing No. 11 of the lower column, forming a seat for brackets 8 and 9, through which the bolts 4 and 5 pass.

8, 9—Are brackets setting on cap 7, extending to the under side of crown-plate 2, with bolts 4 and 5 passing through same, making a bracing support for crown-plate 2, on which beams rest.

10—Represents a reinforcement of steel embedded in concrete of lower column, passing from under side of cap 7 to base of column.

11—Represents steel shell of lower column.

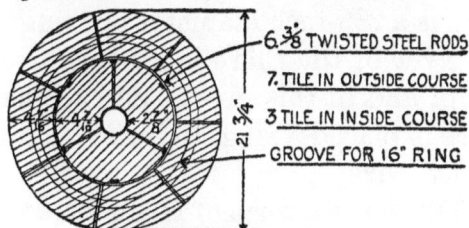

Fig. 111. Terra-Cotta Tile Column.

Fig. 111 shows a form of **terra-cotta tile column.** The column whose cross-section is shown was built of special-shaped, hard-burned, terra-cotta tile, laid with ¼-inch Portland cement mortar joints, and is reinforced by six ⅜-inch twisted steel rods. The special-shaped tile consisted of two concentric rings, the inner one being composed of three tile and the outer one of

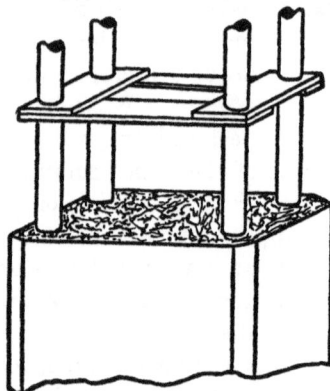

Fig. 112.   Hennebique Column.

seven tile. The reinforcing rods are in the joints between the inner and outer rings. This particular column was 21¾ inches in diameter, and 21 feet 9½ inches high. Through its center, there is a 2⅞-inch diameter opening. In the outside layer of tile there is a 1-inch groove, ¼ inch deep; and in this groove on each course of tile, there is placed a ring, 16 inches in diameter, of $^3/_{16}$-inch wire.

Fig. 112 shows a form of **Hennebique column.**

Plate 19 shows the type of reinforcement used in the columns of the **Turner Mushroom System.** This column was used in a stock-house for the Hamm Brewing Company of St. Paul, and was designed to carry a 1,000-ton load. The finished column was a 30-inch octagon 26 feet high.

REINFORCED CONCRETE WALLS

Walls for buildings may be of various types, largely depending upon the style of structure and the use to which the building is to be put. In the monolithic form of reinforced concrete building, the space between the outside beams, girders, and columns on the ends and sides may be filled in with a curtain wall of brick, tile, or even a reinforced concrete slab with bars running two ways for strength and to prevent shrinkage. In residence work, walls are sometimes built double, with a 4-inch air-space between the two reinforced slabs. In this type of wall, the reinforcing rods should also run horizontally and vertically.

The thickness of slab and amount of reinforcement necessary for a **curtain wall,** or a vertical wall which is to bear no weight, is determined by figuring it as a flat slab supported at all four sides, and carrying a uniformly distributed load of 40 lbs. per square foot due to wind pressure. An ordinary slab designed on this basis will probably be four or five inches thick, allowing for a good factor of safety.

In small structures, it is common practice to build the walls as one continuous reinforced piece, hollow or solid as desired, the forms being carried up as the wall progresses. The reinforcement in such walls consists of lateral rods, wires, expanded metal, or wire fabric, as thought best in the individual case. Window- and door-frames should be thoroughly reinforced to prevent the formation of cracks from one to another.

Metal ties used in hollow walls should have their surfaces covered with cement to prevent corrosion and gradual wasting away. At the corners of the building, the reinforcing rods or material used, from the two sides, should lap over each other so as to make a firm corner joint

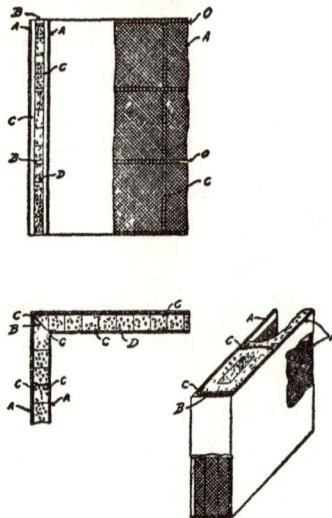

Fig. 113. Lea's Concrete Metal Wall Construction.

and tie the two walls together. This applies to both double and solid walls.

Fig. 113 shows a form of wall construction known as Lea's Concrete Metal Wall. In the figure,

DOUBLE HOUSE—STUCCO ON METAL LATH.

ROW OF DOUBLE HOUSES—STUCCO ON METAL LATH.

A=Wire Fabric.
B=Spacing Bar.
C=Vertical Member.
D=Separator.
O=Horizontal Member.

A frame of the desired form is erected of structural steel and covered with wire fabric as shown. A coating of cement or mortar is then applied to the outside of the wire fabric which, upon hardening, forms a shell of the desired outline, which may be filled in with concrete. This method of construction does not require the use of forms or moulds, thus effecting a great saving in material and labor, besides affording a strong, well-finished structure. It may be employed in building dams, retaining walls, culverts and other structures.

CEILINGS

Fig. 96 also shows the method of supporting a ceiling made of metal lath or sheathing. This sheet material is wired securely to the light bars shown running transversely across the I-beams. The bars are held to the flanges of the I-beams by clamps, the top part of the clamp being bent over the flange and embedded in the concrete. Plaster is placed directly on the metal lath. This same method of fireproofing is also applied to the rafters.

In the case of the Vaughan system shown in Fig. 79, the lower part of the concrete beams forms the ceiling. These may be finished by the application of a single coat of rough plaster, and then the final finishing coat.

Fig. 102 shows the plaster applied directly to the under side of floor-slabs. These slabs are generally of sufficient roughness from the forms so that they will hold the plaster.

In this same figure is illustrated a ceiling formed by a smooth floor-slab. Care in manipulation of the concrete, together with well-made forms of planed boards, tongued-and-grooved and well jointed, will produce a fairly smooth surface. The concrete must be wet, and well worked over with a shovel or hoe when it is placed on the forms, in order to get the larger parts of the aggregate away from the boards, and allow the mortar to sink down and form a smooth face to the mass. Figs. 97 and 98 show the method of forming ceilings for **arched floors** where some form of close-mesh sheet fabric is used for a reinforcing agent. In Fig. 98, the method of protecting the bottom flanges of I-beams from fire is shown. The rib-lath or expanded metal is placed about the bottom flange as a kind of cage, fastened in place securely, and plastered over with a thick coating of cement plaster. As before stated in an earlier chapter, the steel reinforcing materials should be embedded from $\frac{3}{4}$ to $2\frac{1}{2}$ inches, depending upon the importance of the member and the likelihood of a hot fire in the room.

PARTITIONS

Partitions in reinforced concrete construction work consist of both **solid** and **hollow** types. In factory work, partitions may be made of reinforced concrete four inches thick, of tile, or of concrete blocks. For solid partition walls and elevator wells, it is convenient to pour the concrete after the floors are laid. This may be done by leaving a slot in the floor at the proposed location for the partition.

In the plant of the Bush Terminal Company in South Brooklyn, N. Y., the reinforcement in their solid partitions around the elevator and stair wells consisted of ⅜-inch round rods spaced 15 inches apart both horizontally and vertically.

In the Lynn Storage Warehouse, at Lynn, Mass., around the elevators and stairs, and also to enclose the offices on the first floor and storage rooms on the fifth floor, expanded metal partitions were employed. Expanded metal lathing, No. 24 gauge, was wired to 1-inch channel bars placed vertically 12 inches on centers, and the lathing then plastered with five coats so as to form a solid partition 2 inches thick.

The first or scratch coat consisted of one part cement to 3 parts of lime, with the usual quantity of sand and hair. This pressed through the lathing, so that it could be plastered on both sides with a brown coat of lime and cement mortar in proportions 1 part cement to 3 parts of lime mortar, and followed by a finishing coat of the same mortar on both sides.

Solid partitions are built up with two face forms supported against a rigid framework of uprights. In the middle of the space between the face forms, the reinforcing agent is held firmly in place for the concrete. Such partitions run from 3 to 6 inches in thickness, depending upon the location and class of building. Only light aggregates should be used in this class of work.

Hollow partitions are studded, and metal lathing of some form is held by metal fastenings

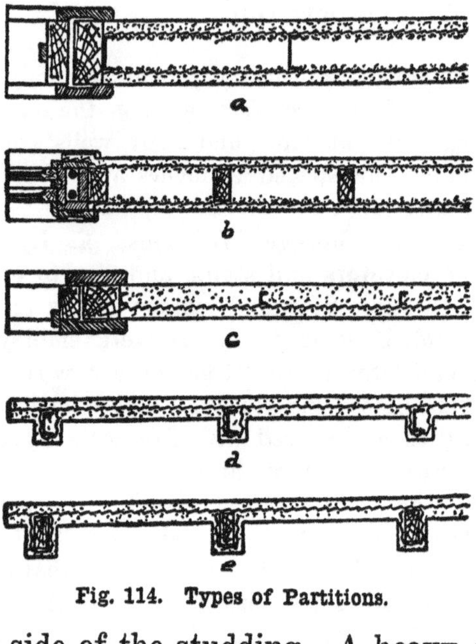

Fig. 114. Types of Partitions.

on each side of the studding. A heavy coat of mortar is then plastered on each side of the metal wall thus formed.

If expanded metal is used as a reinforcing agent in partitions, always lay the metal with the **length of the diamond across the shortest span;** it has only half the strength when placed the other way.

Fig. 114 shows various forms of partitions in which expanded metal fabric is used as reinforcement with Portland cement mortar, with or without lime.

Plate 15 shows a form of partition or light wall construction in which ribbed metal lath is used to hold the plaster. The uprights in this case are of special form used in the Kahn system.

Table XIV gives the standard sizes and gauges of expanded metal meshes and lathing as adopted by the Associated Expanded Metal Companies.

FINISH ON CONCRETE WORK

Floors.—The most common **finish for floors is** the ordinary cement finish. This is a cement mortar composed of one part Portland cement and two parts clean, sharp sand. It is preferably laid at the same time as the main body of the concrete work in order to procure adhesion to the same. If for any reason this cannot be done, the old concrete should be thoroughly cleaned before the finish is laid, and the finish should be made at least one inch in thickness, preferably more. A less thickness will crack off. The cement finish should be marked off in squares, the lines of the marking being so arranged as to bring them over all beams and girders.

TABLE XIV

Standard Sizes and Gauges of Expanded Metal Meshes

Mesh	Gauge (Stubs)	Strand Standard or Extra	Section in Sq. Inches per Foot of Width	Weight per Square Foot, in Pounds	Size of Standard Sheets	Number of Sheets in a Bundle	Number of Sq. Feet in Bundle of 8' 0" Length
¾ in.	No. 18	Standard	.209	.74	4 ft. or 5 ft. x 8 ft.	5	
¾ in.	" 13	"	.225	.80	6 ft. x 8 ft. or 12 ft.	5	240
1½ in.	" 12	"	.207	.70	4 ft. x 8 ft. or 12 ft.	5	160
2 in.	" 12	"	.166	.56	5 ft. x 8 ft. or 12 ft.	5	200
3 in.	" 16	"	.083	.28	6 ft. x 8 ft. or 12 ft.	10	480
3 in.	" 10	Light	.148	.50	6 in. x 8 ft. or 12 ft.	5	240
3 in.	" 10	Standard	.178	.60	6 ft x 8 ft. or 12 ft.	5	240
3 in.	" 10	Heavy	.267	.90	4 ft. x 8 ft. or 12 ft.	5	160
3 in.	" 10	Ex Heavy	.356	1.20	6 ft. x 8 ft. or 12 ft.	3	144
3 in.	" 6	Standard	.400	1.38	5 ft. x 8 ft. or 12 ft.	3	120
3 in.	" 6	Heavy	.600	2.07	5 ft. x 8 ft. or 12 ft.	3	120
4 in.	" 16	Old Style	.093	.42	4½ ft. x 8 ft. or 9 ft.	6	216
6 in.	" 4	Standard	.245	.84	5 ft. x 8 ft. or 12 ft.	5	200
6 in.	" 4	Heavy	.368	1.26	5 ft. x 8 ft. or 12 ft.	3	120

Lathing

Designation	Gauge U S. Standard	Size of Sheets	Sheets in a Bundle	Sq. Yards in a Bundl	Weight per Sq. Yard
A	24	18 x 96	9	12	4½ lbs.
B	27	18 x 96	9	12	3 "
Special B	27	20¼ x 96	9	1.½	2½ "
Diamond No. 24	24	22½ x 96	9	15	3 "
Diamond No. 26	26	24 x 96	9	16	2¾ "

Another method of finishing old floors which is sometimes used is as follows:

A 3-inch layer of 1:2:6 cinder concrete is laid on the old floor, and then the 1:2 coating of cement and sand laid on that. The cinder concrete provides a good bonding surface for the finishing coat and is said to prevent cracking in same.

In factories, hotels, office buildings, etc., where finished wood floors are laid on concrete, beveled wood sleepers are used as nailing strips. These sleepers are about 2 by 3 inches in size, and are placed usually 16 inches on centers. Between the sleepers a filling of weak cinder concrete is used to hold them in place.

Marble, tile, mosaic, and similar floors are laid on concrete construction by imbedding them in a cement mortar.

Walls.—Where a cement finish is desired on concrete walls, the finish should be placed while the wall is being built. The rough concrete is spaded back from the forms, and the rich mortar placed in front of it. A cement finish plastered on concrete after the wall is built will usually crack and not give the best results. After the forms are removed, the concrete should be rubbed smooth and given a coat of cement wash mixed and applied as a paint.

In cases where it is necessary to plaster concrete walls, precaution should be taken to make the plaster stick. In many cases it is well to wash or scrub the surface, or to pick it to make it rough, before applying the plaster. A rich mortar of 1 part cement to from 1 to 2 parts sand should be used for such work. Lime paste added to the mortar is advisable in some cases. This increases the adhesion and lessens the liability of cracking. If a hard surface is desired, only a small amount of the paste should be used. In plaster work of this kind it is customary to brush over the surface after troweling, to remove the tension in the cement.

An inexpensive manner of plastering is what is called a **splatter-dash coat**. The mortar is thrown against the surface with a stick or paddle, making a very effective rough surface. A rough surface is generally better in appearance and less liable to crack than a smooth surface.

A method recommended as providing a good bond is to first wash the surface thoroughly with water, and then brush on a coat of neat cement. While this is wet, a coat of plaster about $\frac{1}{4}$ inch thick is put on. This is followed by succeeding coats applied about an hour apart until the desired thickness is reached. If desired, the last coat may be thrown on in order to produce a rough surface.

Mr. Ernest McCullough advises the following:

"If the appearance of the work requires a coat of plaster, clean the surface with steam, afterwards using wire brushes and then the steam again. Wet it with water, paint it with neat cement, and immediately follow with two coats of 1:3 mortar, the lower coat scratched and the top coat floated to a sand surface."

Plastering should as a general rule be resorted to only to fill holes and to smooth over rough places. Godfrey states that a plaster coat should either be very thin—that is, just enough to fill irregularities—or it should be 1 to 3 inches thick, so that it will have some strength in itself.

Many types and methods of finish are used

on walls—such as placing a rich mortar in front of the concrete when the forms are being filled; or a granolithic surfacing composed of 1 part cement, 2 parts coarse sand or gravel, and 2 parts granolithic grit made into a stiff mortar and placed in a layer about 1 inch thick in front of the concrete in the moulds. The face of the form is removed in this latter form as soon as the concrete has become rigid enough—generally on the following day—and the wall scrubbed with water until the grit shows up. The wall should then be protected and kept moist for three or four days.

Strips of wood are often nailed into the forms to give the effect of cut stone to the surface when the forms are removed. Dry surface finish is produced by using a fine stone in the aggregate, mixing fairly dry and not spading the concrete next to the forms. Facings of brick, tile, stone, and even cast slabs of concrete, are used in walls for a finish, backed up by the rougher poured work.

There are many other ways of obtaining pleasing appearances to finished concrete work, such as **bush-hammering, pebble dash, Quimby process,** and a large variety of patented processes, all of which have been used with more or less success.

INDEX

A

PAGE
Acid Treatment of Concrete
Surfaces 35
Aggregates, Colored....... 42
Aiken Method of House
Construction 84
Alignment of Forms....... 11
Alligater Point on Pile.... 283
"American" System of Re-
inforcement 273, 307
Anchored End Bar Rein-
forcement 310
Apron 140
Arched Floors 304, 322
Asbestos Building Lumber. 43
Asbestos Shingles 50
Axis, Neutral 183, 220

B

Barbed Wire 243
Barns and Stables......... 127
Bars 221
Bent or Twisted........ 248
Deformed 238
Plain 238
Steel 238
Trussed 238
Barton "Spider-Web" Sys-
tem of Reinforcement
258, 307
Battered Walls 28
Beams, Concrete, Stresses in
184, 231
Bending Moment 225
Bent or Twisted Bars...... 248
Black Stone 91
Blaw Centering 23
Blocks, Concrete—
Curing 99
Industry 93
Making 102
Proportions for Mixing.. 96
Use of 94
Cement for 96
Color for 90
For Foundation Walls... 32
Without Facing 86
Blue Stone 92
Box Stalls 130

PAGE
Brick, Cement, for Chim-
neys 46
Brown or Buff Stone....... 92
Building Construction, Gen-
eral 296
Building Laws of Different
Cities 235, 236
Bush-Hammering 35, 329

C

Cast Steps 61
Causes of Defects......... 174
Causes of Failure.......... 15
Ceilings 234, 321
Cellar—
Cyclone 151
Floors 53
Mushroom 151
Root 149
Cement Blocks—
Curing 99
Making 102
Proportions for Mixing.. 96
Use of 94
(See also Blocks, Concrete.)
Cement Brick for Chimneys 46
Cement Coping 65
Cement Floors and Steps... 51
Cement for Blocks......... 96
Cement Roofing 46
Centering, Blaw 23
Centering, Cost of......... 203
Chair, National 278
Chenoweth Concrete Piles.. 291
Chicken House 143
Chimney Caps 67
Chimneys, Cement Brick for 46
Chimneys, Roofs, etc....... 45
Cinder-Concrete Roof 314
Cinders, Use of, in Sidewalk
Construction 170
Circular Forms 20
Cisterns 110
Cities, Building Laws of.235, 236
Colored Aggregates 42
Color of Concrete Blocks... 90
Color Variations 18
Collapsible Forms 138

PAGE

Columbian System of Rein-
    forcement 254
Column Footings, Reinforced 281
Column Hoops 241, 316
Columns, Reinforced Con-
    crete 231, 314
Compression 183, 231
Concrete Block Industry... 93
Concrete Blocks—
    Color for 90
    For Foundation Walls.... 32
    (See also Blocks, Concrete.)
Concrete on the Farm...... 108
Concrete Piazza 64
Concrete Pile Foundations.
                       209, 282
Concrete, Reinforced...... 183
    Advantages of 193
    Cost of 198
    Design of 220
    History of 189
Concrete Roofs 45
Concrete Steps in Damp
    Places 57
Concrete, Weight of....... 8
Coping, Cement 65
Construction, Form 1, 20
Construction, General Build-
    ing 296
Contraction in Walls...... 30
Corners, Protection of..... 178
Corrosion of Steel........ 249
Cost of Labor.......4, 202, 203
Cost of Reinforced Concrete
                       198, 207
Cost of Various Classes of
    Concrete Work........ 207
Cost of Walls...........33, 207
Cost of Wood and Concrete
    Piles 209
Cowles "Umbrella" Sys-
    tem of Reinforcement.
                       262, 307
Cracks in Concrete........ 175
Culvert Construction...137, 139
Culvert Forms 21
Culverts, Road 137, 139
Cummings System of Rein-
    forcement 254
Curbs and Cutters........ 176
Curing Cement Blocks..... 99
    Time for 100
Curtain Walls 319

PAGE

Cutting with Hammer..... 41
Cyclone Cellar 151

D

Defects, Causes of........ 174
Deformed Bars 238
Design of Forms........... 5
Design of Reinforced Con-
    crete 220
De Vallière System........ 268
Diamond Bar 239
Diamond Cement Shingles.. 48
Durability 197

E

Edison Poured House...... 80
English Systems of Floor
    Construction 70, 325
Erection   of   Reinforced
    Structures 299
Excavating 31
Expanded Metal...242, 243, 326
Expansion Joints for Side-
    walks 171

F

Factories,   Concrete,   vs.
    Wood or Brick........ 217
Failure, Causes of, in Con-
    crete Work 15
Farm, Concrete on the..... 108
Fawcett's Floor 75
Feeding Floors........ ... 129
Fence-Posts, Concrete 151
    Reinforced 154
Ferguson's Floor 74
Ferroinclave 311
Ferro-Lithic Plate for Roof-
    ing 49
Finish for Monolithic Struc-
    tures 19
Finishes for Concrete Work
                    33, 325, 327
Fireplace, Concrete 69
Fireproofing 70, 76
Fire-Resisting Properties of
    Concrete 211
Fire Risk and Insurance.214, 217
Flat Slabs 303, 322
Floor Construction, English
    Systems 70, 325
Floor Loads, Allowable..... 235
Floors and Roofs, Materials
    for 234, 302

PAGE

Floors and Steps of Cement   51
Floors, Barn and Stable.... 127
    Cellar   53
    Feeding 129
    Finish for 325
    Girderless 308
    Jointless   52
Floors, Slabs, and Roofs...
                    234, 302, 322
Flying Stairs or Steps.....   54
Floors, Veranda   52
Footings 28, 281
Forms—
    Alignment of   11
    Circular   20
    Collapsible 138
    Construction 1, 20
    Cost of 203
    Culvert   21
    Design of   5
    Metal   22
    Metal Collapsible Type..   22
    Protection of   12
    Strength of 9, 17
    Symmetry of   17
    Time to Remove........   13
Formulas for Design—
    Ransome's 227
    Wason's 225
Foundations   28
    House   30
    Pile 282
Fountain 135
Frame Construction 297
Frames, Hotbed 133
Frazzi Floor   76
Frozen Concrete 174

G

Gabriel System of Rein-
    forcement........252, 305
Gate-Posts, to Preserve.... 155
Gilbreth Corrugated Piles. 289
Girderless Floor Construc-
    tion 308
Gow Concrete Piles....... 287
Graham System of House
    Construction   81
Greenhouses 133
Grooved Surfaces 130
Grumman Cement Shingles.   48
Gutters 176

H

PAGE

Hammer, Cutting with.....   41
Hardening of Concrete.... 208
Hearth, Concrete   69
Heat, Effect of.........79, 211
Hennebique Column 318
Hennebique Concrete Piles. 295
Hennebique System of Re-
    inforcement......268, 305
Hennebique Trussed Bar.. 240
Hens' Nests, Concrete..... 144
Herringbone Expanded Steel
    Lath 242
Herringbone Trussed Bar.
                      263, 265
High-Carbon Steel 273
History of Reinforced Con-
    crete 189
Hog Pens 145
Hollow-Wall Method of Con-
    struction   83
Hollow Walls   28
Horse Block, Concrete.... 131
Hotbed Frames 133
House Construction, Con-
    crete   80
House Foundations   30

I

Ice-House 147
Independent Bar Reinforce-
    ment 303
Insurance, Cost of.........217

J

Jackson System of House
    Construction   28
Johnson Bar 239
Jointless Floors   52
Joist Supports   32

K

Kahn System of Reinforce-
    ment 251, 304, 307
Kahn Trussed Bar.....239, 251
Kleine Floor   75

L

Labor Cost   4
Lagging 27, 140
Lally Column 316
Lath, Metal 42, 243
    Standard Sizes and
        Gauges 326

PAGE

Lattice Construction 653
Latticed Reinforcement... 303
Laws Regarding Reinforced
   Concrete 235, 236
Lea's Concrete Metal Wall. 320
Lettering with Cement.... 102
Life of Steel in Concrete.. 249
Loads, Allowable, on Floors 235
Lock-Bar Steel-Concrete
   Frame 266
Lumber, Finish and Thick-
   ness of 10
Lumber, Kinds to Use for
   Forms 9

M

Making Slabs 63
Making Cement Blocks.... 102
Materials for Floors and
   Roofs 234
Mason, Problems for the... 76
Materials, Reinforcing 238
"Mercantile" Pipe 143
Merrick Floor System.270, 304
Metal for Reinforcement.. 238
Metal Forms 22
Metal Forms, Collapsible
   Type 22
Metal Lath 42, 243
Mill Construction 297
Miracle Moulds 21
Mixtures for Various Class-
   es of Work........... 301
Monolith Steel Bar........ 238
Monolith System 23
Mortar, Coloring 91
Mortar, Effect of Heat on.. 79
Mortar, Hardness and Re-
   sistance of 80
Mortar, Strength of...... 101
Mould, Miracle 21
Mould, Overturf 23
Moulds for Ornaments..... 157
Multiplex Steel Plate...... 241
Mushroom Cellar 151
"Mushroom" System of Re-
   inforcement...256, 307, 318

N

National System of Rein-
   forcement 278, 304
Nests, Hens'.............. 144
Neutral Axis 183, 220

O

PAGE

Oil, Use of, on Forms..... 11
Ornaments, Moulds for.... 155
Overturf Mould 23
Over-Troweling 175

P

Painting Concrete Surfaces. 42
Parabolic Theory 224
Partitions 31, 323
   Hollow 324
   Solid 323
Pebble-Dash Finish 329
Piazza, Concrete 64
Piers and Posts........... 131
Pile Foundations, Concrete. 282
Piles—
   Chenoweth 291
   Gilbreth Corrugated 289
   Gow 287
   Hennebique 295
   Raymond 285
   Simplex 283
Piles, Concrete and Wood,
   Cost of 209
Pin-Connected Girder Frame 265
Pipe, "Mercantile"....... 143
Pipe, Sewer 141
Plain Bars 238
Platform Steps 60
Plumb and Battered Walls. 28
Pock Marks in Concrete... 175
Porch Steps 60
Porosity, Absence of...... 298
Posts, Concrete 131
Poultry House 143, 144
Problems for the Mason... 76
Proportions 31
Protecting Forms 12

Q

Quimby Finish for Concrete
   Walls 329

R

Rain Barrel 132
Ransome Bar 239
Ransome's Formula 227
Ransome System of Rein-
   forcement 250
Raymond Concrete Piles... 285
Red Stone 92
Reinforced Concrete 183

	PAGE
Advantages of	193
Building Laws Regarding	235, 236
Cost of	198
Design of	220
Durability of	197
General Principles	183
History of	189
Stiffness of	198
Reinforcing Materials and Systems	238, 250
Reinforcement for Fence-Posts	154
Reinforcement, Systems of.	250
Resisting Moment	255
Rib-Metal	242, 310
Road Culverts	137
Rods or Bars	221
Roebling Floor System.	270, 304
Roof, Cinder-Concrete	314
Roofing, Cement	46
Roofs, Concrete.	45, 302, 311, 313
Tile	47, 313
Root Cellar	149
Rubbing Down	19

S

Safe, Concrete	106
Sand Analysis	167
Sarco-Mastic	51
Scrap Iron	243
Scrubbing Concrete Surfaces	35
Setting and Hardening of Concrete	208
Sewer Pipe, Concrete	141
Sheet Fabrics, Metal	221
Shearing Forces	185
Sheet Metal	243
Shingles, Asbestos	50
Concrete	45, 48
Short-Span Floor-Slabs	309
Sidewalk Construction	160
Sidewalks—	
Aggregates	162
Cement	162
Defects in	174
Drainage	168
Expansion Joints	171
Forms	170
Foundation or Sub-Base.	168
Quantity of Material Required	174

	PAGE
Surface Treatment	172
Side Walls for Steps.	61
Siegwart System of Floor Construction	273
Silos, Concrete, Advantages of	120
Concrete Block	125
Construction of	126
Hollow-Wall and Monolithic	118
Kinds of	122
Sizes and Capacities of.	122
Simplex Piles	283
Sinks	109
Slabs, Floor and Roof	302
Flat and Arched	303
Stair	63
Slipping of Steel in Concrete	185
Sloughing Off from Concrete Surfaces	175
Slow-Burning Construction.	297
Smith Girderless Floor Construction	308
Smith Spiral Column	315
"Spider-Web" System.	258, 307
Spiral Column, Smith	315
Splatter-Dash Finish	328
Square Lug Bar	239
Stables	127
Stairs or Steps, Flying	54
Stalls, Box	130
"Standard" System of Floor Construction	272
Steel Bars for Reinforcement	238
Steel Construction	297
Steps—	
Concrete	51
In Damp Places	57
Cast	61
For Residences	55
On Terraced Grounds	58
Platform	60
Porch	60
Side Walls for	61
Stiffness of Reinforced Concrete Structures	198
Stirrups	186
Stone, Black	91
Blue	92
Brown or Buff	92
Red	92
Storage Buildings	148

PAGE

Straight-Line Theory 224
Strength of Forms 17
Structural Shapes 221
Stucco 39
Sunburned Surface 174
"Superior" Cement Shingles 49
Symmetry of Forms....... 17
System "M"............. 266
Systems of Reinforcement.
                    238, 250

T

Tamping Concrete 175
Tanks and Cisterns........ 110
T-Beams 210
Tensile Stresses 185
Tension 183
Terra - Cotta Lining for
    Walls 32
Terra-Cotta Tile Column... 317
"Thomas" Wall 29
Tile, Concrete 179
    Ageing 182
    Cost of 179
    Freezing of 182
    Proportions for 179
Tile Roof 47, 313
Tooling Concrete Surfaces. 35
Trussed Bars 238
Trussit 242
Turner "Mushroom" Sys-
    tem 256, 307, 318
Twisting Bars 248

U

"Umbrella" System,
    Cowles 262, 307
Unit-Beam System of Re-
    inforcement 265

PAGE

Unit-Frames 221
Unit-Girder Frame.263, 265, 307
Unit of Measurement...... 96
U-Twisted Lock Bar....... 18

V

Variations in Color....... 18
Vaughan System of Floor
    Construction 275, 304
Veranda Floors 52

W

Wainwright System for Pro-
    tecting Corners 178
Walls—
    Contraction in 30
    Cost of 33
    Finishes for 33, 325, 327
    Foundation 28
    Foundation, Concrete
        Blocks for 32
    Lea's Concrete Metal.... 320
    Partition 31
    Plumb and Battered..... 28
    Reinforced Concrete 319
    Solid and Hollow....... 28
Wason's Formula 225
Waterproof Concrete Tank. 112
Water-Tanks 110
Weight of Concrete....... 7
Well Curbs 117
Windmill Foundation 109
Wire—
    For Reinforcement 243
    Barbed 243
Wire Fabric 242
Wire Rope or Cable....... 244
Woven Wire 243